USING SI UNITS IN ASTRONOMY

A multitude of measurement units exist within astronomy, some of which are unique to the subject, causing discrepancies that are particularly apparent when astronomers collaborate with other disciplines in science and engineering. The International System of Units (SI) is based on a set of seven fundamental units from which other units may be derived. However, many astronomers are reluctant to drop their old and familiar systems. This handbook demonstrates the ease with which transformations from old units to SI units may be made. Using worked examples, the author argues that astronomers would benefit greatly if the reporting of astronomical research and the sharing of data were standardized to SI units. Each chapter reviews a different SI base unit, clarifying the connection between these units and those currently favoured by astronomers. This is an essential reference for all researchers in astronomy and astrophysics, and will also appeal to advanced students.

RICHARD DODD has spent much of his astronomical career in New Zealand, including serving as Director of Carter Observatory, Wellington, and as an Honorary Lecturer in Physics at Victoria University of Wellington. Dr Dodd is Past President of the Royal Astronomical Society of New Zealand.

USING SI UNITS IN ASTRONOMY

RICHARD DODD

Victoria University of Wellington

Shaftesbury Road, Cambridge CB2 8EA, United Kingdom

One Liberty Plaza, 20th Floor, New York, NY 10006, USA

477 Williamstown Road, Port Melbourne, VIC 3207, Australia

314–321, 3rd Floor, Plot 3, Splendor Forum, Jasola District Centre, New Delhi – 110025, India

103 Penang Road, #05–06/07, Visioncrest Commercial, Singapore 238467

Cambridge University Press is part of Cambridge University Press & Assessment,
a department of the University of Cambridge.

We share the University's mission to contribute to society through the pursuit of
education, learning and research at the highest international levels of excellence.

www.cambridge.org
Information on this title: www.cambridge.org/9780521769174

First published 2012

A catalogue record for this publication is available from the British Library

Library of Congress Cataloging-in-Publication data
Dodd, Richard.
Using SI units in astronomy / Richard Dodd.
p. cm.
Includes bibliographical references and index.
ISBN 978-0-521-76917-4 (hardback)
1. Communication in astronomy. 2. Metric system. I. Title.
QB14.2.D63 2011
522'.87–dc23 2011038728

ISBN 978-0-521-76917-4 Hardback

Contents

Preface

Other than derogatory comments made by colleagues in university physics departments on the strange non-standard units that astronomers used, my first unpleasant experience involved the *Catalog of Infrared Observations* published by *NASA* (Gezari *et al.*, 1993). In the introduction, a table is given of the 26 different flux units used in the original publications from which the catalogue was compiled – no attempt was made to unify the flux measures. The difficulties of many different ways of expressing absolute and apparent flux measures when trying to combine observations made in different parts of the electromagnetic spectrum became all too apparent to me when preparing a paper (Dodd, 2007) for a conference on standardizing photometric, spectrophotometric and polarimetric observations. This work involved plotting X-ray, ultraviolet, visible, infrared and radio frequency measurements of selected bright stars in the open cluster IC2391 as spectra with common abscissae and ordinates. Several participants at the conference asked if I could prepare a 'credit card' sized data sheet containing the conversion expressions I had derived. As is usually the case, I was otherwise engaged at the time in comparing my newly derived coarse spectrophotometry with a set of model stellar atmospheres, so the 'credit card' idea was not acted upon. However, the positive response to my paper did make me realize that there was a need in the astronomical community for a reference work which, at the least, converted all the common astronomical measurements to a standard set. The answer to the question 'Which set?' is fairly self evident since it was over 40 years ago that scientists agreed upon a metric set of units (Le Système International d'Unités or SI units) based on three basic quantities. For mass there is the kilogram, for length the metre and for time the second. This primary group is augmented by the ampere for electric current, the candela for luminous intensity, the kelvin for temperature and the mole for amount of substance. From these seven it is relatively easy to construct appropriate physical units for any occasion: e.g., the watt for power, the joule for energy or work, the newton for force and the tesla for magnetic flux density.

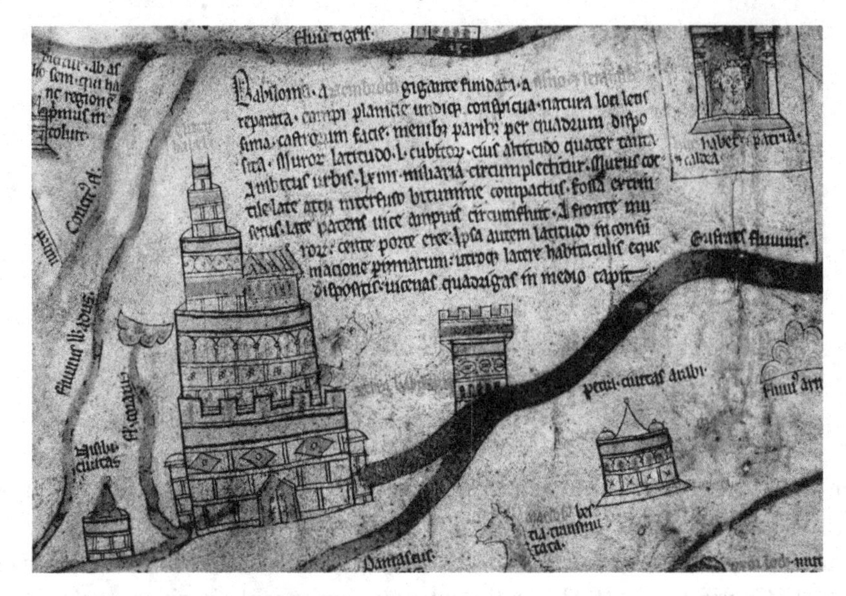

The thirteenth-century *Mappa Mundi* illustration of the Tower of Babel. (© The Dean and Chapter of Hereford Cathedral and the Hereford Mappa Mundi Trust.)

It is possible to express even the more unusual astronomical quantities in SI units. The astronomical unit, the light year and the parsec are all multiples of the metre – admittedly very large, and non-integral, multiples from $\sim$150 billion metres for the astronomical unit to $\sim$31 quadrillion metres for the parsec. Similarly, one solar mass is equivalent to $\sim 2 \times 10^{30}$ kilograms and the Julian year (365.25 days) to $\sim$31.6 million seconds. So in each of these cases we could use SI units, though quite obviously many are unwieldy and a good scientific argument for using special astronomical units may readily be made.

In many areas of astronomy, the combination of research workers trained initially at different times, in different places and in different disciplines (physics, chemistry, electrical engineering, mathematics, astronomy etc.) has created a Babel[1]-like situation with multitudes of units being used to describe the same quantities to the confusion of all.

Astronomers participate in one of the most exciting and dynamic sciences and should make an effort to ensure the results of their researches are more readily available to those interested who may be working not only in other branches of astronomy but also in other fields of science. This can be done most readily by using the internationally agreed sets of units.

[1] The story of the Tower of Babel is set out in the Hebrew Bible in the book of Genesis, chapter 11, verses 1–9, and relates to problems caused by a displeased God introducing the use of several different, rather than one spoken language, to the confusion of over-ambitious mankind. A depiction of the Tower of Babel that appears on the thirteenth-century *Mappa Mundi* in Hereford Cathedral Library is shown in the reproduction above.

However, stern, but sensible, comments from the reviewers of the outline of the proposed book, plus a great deal more reading of relevant astronomical texts on my part, has led to a better understanding of why some astronomers would be reluctant to move away from non-standard units. This applies particularly in the field of celestial mechanics and stellar dynamics, where the International Astronomical Union approved units include the astronomical unit and the solar mass. However, this in itself should not act as a deterrent from adding SI-based units alongside the special unit used, with suitable error estimates to illustrate why the special unit is necessary.

In a recent book review in *The Observatory*, Trimble (2010) admitted to appending an average of about two corrections and amplifications per page in not only the book she had just reviewed but also in her own book on stellar interiors (Hansen *et al.*, 2004), and Menzel (1960) completed the preface to his comprehensive work *Fundamental Formulas of Physics* by stating: 'In a work of this magnitude, some errors will have inevitably crept in.'

Whilst, naturally, I hope that this particular volume is flawless, I must confess I consider that to be unlikely! The detection and reporting of mistakes would prove of considerable value and, likewise, comments from readers and users of the book on areas in which they believe it could be improved would be welcome. My own experience using various well-known reference works and textbooks, to some of which I had previously assigned an impossibility of error, was that they all contained mistakes; some travelled uncorrected from one edition to the next and others in which correct numerical values or terms in an algebraic equation in an earlier edition were incorrectly transcribed to a later.

The most radical suggestions in this book are probably: a simple way of describing and dealing with very large and very small numbers; the use of a number pair of radians rather than a combination of three time and three angular measures to locate the position of an astronomical body; and the replacement of the current ordinal relative-magnitude scheme for assigning the brightness of astronomical bodies by a cardinal system based on SI units in which the brighter the object the larger the magnitude.

Writing a book such as this takes time. Time during which new values of astronomical and physical constants may become available. I have referenced the various sources of constants published before the end of 2010 that were used in the preparation of tables and in the worked examples presented. Readers are invited to substitute later values for the constants, as a valuable exercise, in the worked examples should they so wish.

In conclusion, it is important to bear in mind that the primary purpose of this book is to act as a guide to the use of SI units in astronomy and not as an astronomical textbook.

Acknowledgements

It is a pleasure to thank the following organizations for permission to reproduce illustrations and text from their material.

The Bureau International des Poids et Mesures (BIPM) for permitting the use of the English translations of the formal definitions for each of the SI units and some of the tabular material contained in the 8th edition of the brochure *The International System of Units (SI)* (BIPM, 2006).[2]

The Dean and Chapter of Hereford Cathedral for permission to use a print of part of the *Mappa Mundi* that shows an imagined view of the Tower of Babel.

The Canon Chancellor of Salisbury Cathedral for permission to use their translation from the Latin of clause 35 of the Magna Carta.

Writing a book such as this has benefitted considerably from the availability of online data sources. Those which were regularly consulted included: the Astrophysical Data Service of NASA; the United States Naval Observatory for astrometric and photometric catalogues; the European Southern Observatory for the Digital Sky Surveys (DSS) and the HIPPARCOS and TYCHO catalogues; SIMBAD for individual stellar data; the Smithsonian Astrophysical Observatory for DS9 image analysis software, and many of the other databases and virtual observatory sites listed in Chapter 12.

At an individual level, the inspiration to start this work is due in part to: Mike Bessel, Ralph Bohlin, Chris Sterken, Martin Cohen and other participants at the Blankenberge conference on standardization who expressed an interest in the paper I presented there. Denis Sullivan of the School of Chemical and Physical Sciences of the Victoria University of Wellington provided the enthusiasm and logistical support to continue with this work, and with Mike Reid was responsible for improving my limited skill with LaTeX. Harvey McGillivray, formally of the Royal Observatory

[2] Please note that theses extracts are reproduced with permission of the BIPM, which retains full internationally protected copyright.

Edinburgh, provided me with COSMOS measuring machine data of the double cluster of galaxies A3266. The desk staff at Victoria University of Wellington library and the librarian of the Martinborough public library were of great assistance in sourcing various books and articles. The proofreading was bravely undertaken by Anne and Eric Dodd. To all these people I express my thanks.

My aim was to write a book that would prove of use to the astronomical community and persuade it to move towards adopting a single set of units for the benefit of all. I hope it succeeds!

Edinburgh provided me with [illegible] a pleasant [illegible] double [illegible] of galaxy [illegible] University of [illegible] and the librarian of the [illegible] library [illegible] books and articles. The [illegible] Allied [illegible] all these people [illegible].

[illegible] a book [illegible] would prove of use to the [illegible] and [illegible] I hope it succeeds.

1
Introduction

1.1 Using SI units in astronomy

The target audience for a book on using SI units in astronomy has to be astronomers who teach and/or carry out astronomical research at universities and government observatories (national or local) or privately run observatories. If this group would willingly accept the advantages to be gained by all astronomers using the same set of units and proceed to lead by example, then it should follow that the next generation of astronomers would be taught using the one set of units. Since many of the writers of popular articles in astronomy have received training in the science, non-technical reviews might then also be written using the one set of units. Given the commitment and competence of today's amateur astronomers and the high-quality astronomical equipment they often possess, it follows that they too would want to use the one set of units when publishing the results of their research.

As to why one set of units should be used, a brief search through recent astronomical literature provides an answer. Consider the many different ways the emergent flux of electromagnetic radiation emitted by celestial bodies and reported in the papers listed below and published since the year 2000, is given.

Józsa *et al.* (2009) derived a **brightness temperature of** $\mathbf{4 \times 10^5}$ **K** for a faint central compact source in the galaxy IC2497 observed at a radio frequency of **1.65 GHz**.

Bohlin & Gilliland (2004), using the Hubble Space Telescope to produce absolute spectrophotometry of the star Vega from the far ultraviolet (**170 nm**) to the infrared (**1010 nm**), plotted their results in $\mathbf{erg\,.\,cm^{-2}\,.\,s^{-1}\,.\,\AA^{-1}}$ flux units.

Broadband **BVRI** photometric observations, listed as magnitudes, were made by Hohle *et al.* (2009) at the University Observatory Jena of OB stars in two nearby, young, open star clusters.

In the study of variable stars in the optical part of the spectrum it is quite common to use **differential magnitudes** where the difference in output flux between the variable object of interest and a standard non-varying star is plotted against time or phase (see, e.g., Yang, 2009).

An X-ray survey carried out by Albacete-Colombo *et al.* (2008) of low-mass stars in the young star cluster Trumpler 16, using the Chandra satellite, gives the median X-ray luminosity in units of $\mathbf{erg\,.\,s^{-1}}$.

The integral γ-ray photon flux above **0.1 GeV** from the pulsar J0205 + 6449 in SNR 3C58, measured with the Fermi gamma-ray space telescope, is given in units of $\mathbf{photons\,.\,cm^{-2}\,.\,s^{-1}}$ by Abdo *et al.* (2009).

These are just a few examples of the many different units used to specify flux. Radio astronomers and infrared astronomers often use janskys ($\mathbf{10^{-26}\,W\,.\,m^2\,.\,Hz^{-1}}$), whilst astronomers working in the ultraviolet part of the electromagnetic spectrum have been known to use flux units such as ($\mathbf{10^{-9}\,erg\,.\,cm^{-2}\,.\,s^{-1}\,.\,\text{Å}^{-1}}$) and ($\mathbf{10^{-14}\,erg\,.\,s^{-1}\,.\,cm^{-2}\,.\,\text{Å}^{-1}}$). So it would seem not unreasonable to conclude that whilst astronomers may well be mindful of SI units and the benefits of unit standardization they do not do much about it.

Among reasons cited in Cardarelli (2003) for using SI units are:

1. It is both metric (based on the metre) and decimal (base 10 numbering system).
2. Prefixes are used for sub-multiples and multiples of the units and fractions eliminated, which simplifies calculations.
3. Each physical quantity has a unique unit.
4. Derived SI units, some of which have their own name, are defined by simple expressions relating two or more base SI units.
5. The SI forms a coherent system by directly linking the mechanical, electrical, nuclear, chemical, thermodynamic and optical units.

A cursory glance at the examples given above shows numerous routes to possible mistakes. Consider the different powers of ten used, especially by ultraviolet astronomers. Some examples use wavelengths, some frequencies, and some energies to define passbands. One uses a form of temperature to record the flux detected. In short, obfuscation on a grand scale, which surely was not in the minds of the astronomers preparing the papers. For this book to prove successful it would need to assist in a movement towards the routine use of SI units by a majority, or at the very least a large minority, of astronomers.

1.2 Layout and structure of the book

The introductory chapter (1) contains the reasons for writing the book and the target audience, definitions of commonly used terms, a brief history of the standardization of scientific units of measurement and a short section on the future of SI units.

Descriptions of the base and common derived SI units, plus acceptable non-SI units and IAU recommended units, are listed in Chapter 2 with Conférence Général des Poids et Mesures (CGPM) approved prefixes and unofficial prefixes for SI units with other possible alternatives.

Given the importance of the technique known as dimensional analysis to the study of units, an entire chapter (3) is allocated to the method, including worked examples. There are further examples throughout the book that illustrate the value of dimensional analysis in checking for consistency when transforming from one set of units to another.

Eight chapters (4–11) cover the seven SI base units plus the derived unit, the radian. Each includes the formal English language definition published by the Bureau International des Poids et Mesures (BIPM) and possible future changes to that definition. Examples of the uses of the unit are given, including transformations from other systems of units to the SI form. Derived units, their definitions, uses and transformations are also covered, with suitable astronomical worked examples provided. Each chapter ends with a summary and a short set of recommendations regarding the use of the SI unit or other International Astronomical Union (IAU) approved astronomical units.

The book ends with a chapter (12) on astronomical taxonomy, outlining various classification methods that are often of a qualitative rather than a quantitative nature (e.g., galaxy morphological typing, visual spectral classification).

The subject matter of the book covers almost all aspects of astronomy but is not intended as a textbook. Rather, it is a useful companion piece for an undergraduate or postgraduate student or research worker in astronomy, whether amateur or professional, and for the writers of popular astronomical articles who wish to link everyday units of measurement with SI units.

1.3 Definitions of terms (lexicological, mathematical and statistical)

The meaning of a word is, unfortunately, often a function of time and location and is prone to misuse, rather as Humpty Dumpty said in *Through the Looking Glass*, 'When I use a word, it means just what I chose it to mean – neither more nor less.'[3]

When discussing a subject such as the standardization of units, it is of paramount importance to define the terms being used. Hence, words that appear regularly throughout the book related to units and/or their standardization are listed in this section with the formal definition, either in their entirety or in part, as given in

[3] See Carroll L. (1965). *Through the Looking Glass*. In *The Works of Lewis Carroll*. London: Paul Hamlyn, p. 174.

volumes I and II of *Funk & Wagnalls New Standard Dictionary of the English Language* (1946).

1.3.1 Lexicological and mathematical

Unit

Any given quantity with which others of the same kind are compared for the purposes of measurement and in terms of which their magnitude is stated; a quantity whose measure is represented by the number 1; specifically in arithmetic, that number itself; unity. The numerical value of a concrete quantity is expressed by stating how many units, or what part or parts of a unit, the quantity contains.

Standard

Any measure of extent, quantity, quality, or value established by general usage and consent; a weight, vessel, instrument, or device sanctioned or used as a definite unit, as a value, dimension, time, or quality, by reference to which other measuring instruments may be constructed and tested or regulated.

The difference between a unit and a standard is that the former is fixed by definition and is independent of physical conditions, whereas a standard, such as the one-metre platinum–iridium rod held at the Bureau International des Poids et Mesures (BIPM) in Sèvres, Paris, is a physical realization of a unit whose length is dependent on physical conditions (e.g., temperature).

Quantity (Specific)

(1) Physics: A property, quality, cause, or result varying in degree and measurable by comparison with a standard of the same kind called a unit, such as length, volume, mass, force or work.
(2) Mathematics: One of a system or series of objects having only such relations, as of number or extension as can be expressed by mathematical symbols; also, the figure or other symbol standard for such an object. Mathematical quantities in general may be real or imaginary, discrete or continuous.

Measurement

The act of measuring; mensuration; hence, computation; determination by judgement or comparison. The ascertained result of measuring; the dimensions, size, capacity, or amount, as determined by measuring.

The mathematical definition of a quantity Q, is the product of a unit U, and a measurement m, i.e.,

$$Q = m\,U \tag{1.1}$$

Q is independent of the unit used to express it. Units may be manipulated as algebraic entities (see Chapter 3) and multiplied and divided.

Dimension

Any measurable extent or magnitude, as of a line, surface, or solid; especially one of the three measurements (length, width and height) by means of which the contents of a cubic body are determined; generally used in the plural. Any quantity, as length, time, or mass, employed or regarded as a fundamental factor in determining the units of other physical quantities (see Chapter 3); as, the dimensions of velocity are length divided by time. The dimension of a physical quantity is the set containing all the units which may be used to express it, e.g., the dimension of mass is the set (kilogram, gram, pound, ton, stone, hundredweight, grain, solar mass ...).

Accurate

Conforming exactly to truth or to a standard; characterized by exactness; free from error or defect; precise; exact; correct.

Accuracy

The state or quality of being accurate; exactness; correctness.

Precise

Having no appreciable error; performing required operations with great exactness.

Precision

The quality or state of being precise; accuracy of limitation, definition, or adjustment.

There is a tendency to use *accuracy* and *precision* as though they had the same meaning, this is not so. Accuracy may be thought of as how close the average value (see below) of the set of measurements is to what may be called the correct or actual value, and precision is a measure (see standard deviation below) of the internal consistency of the set of measurements. So if, for example, a measuring instrument is incorrectly set up so that it introduces a systematic bias in its measurements, these measures may well have a high internal consistency, and hence a high precision, but a low accuracy due to the instrumental bias.

Error

The difference between the actual and the observed or calculated value of a quantity.

Mistake

The act of taking something to be other than it is; an error in action, judgement or perception; a wrong apprehension or opinion; an unintentional wrong act or step; a blunder or fault; an inaccuracy; as a mistake in calculation.

1.3.2 Statistical

Statistics is an extensive branch of mathematics that is regularly used in astronomy (see Wall & Jenkins, 2003). As a very basic introduction to simple statistics, definitions are given for the terms *mean* and *standard deviation*, which are commonly used by astronomers and an illustration (Figure 1.1) of the Gaussian or normal distribution curve showing how a set of random determinations of a measurement are distributed about their mean value.

Mean

Consider a set of N independent measurements of the value of some parameter x then the mean, or average, value, μ, is defined as:

$$\mu = \frac{1}{N}\sum_{i=1}^{N} x_i \tag{1.2}$$

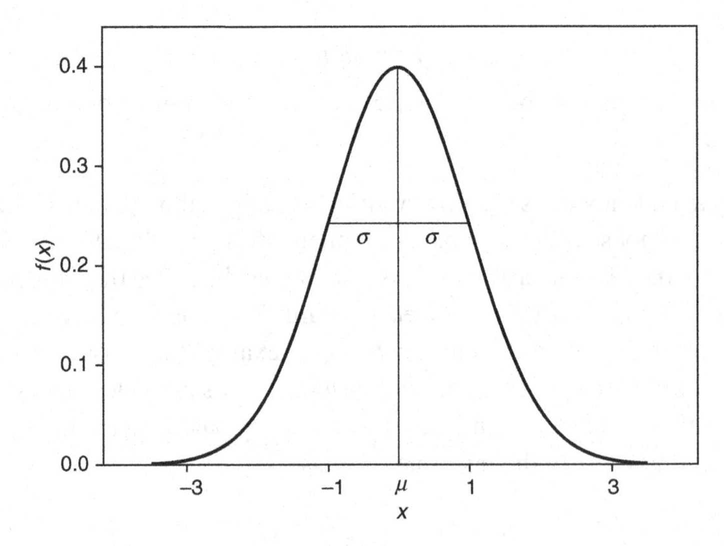

Figure 1.1. A Gaussian distribution with $\mu = 0$ and $\sigma = 1$ generated using equation (1.4).

Standard deviation

Given the same set of N measurements as above, the standard deviation, σ, is defined as:

$$\sigma = \frac{1}{N}\sqrt{\sum_{i=1}^{N}(x_i - \mu)^2} \tag{1.3}$$

Gaussian distribution

For large values of N, the expression describing the Gaussian or Normal distribution of randomly distributed values of x about their mean value μ is:

$$f(x) = \frac{1}{\sigma\sqrt{(2\pi)}} e^{\left(\frac{-1}{2\sigma^2}(x-\mu)^2\right)} \tag{1.4}$$

Figure 1.1 shows the typically bell shaped Gaussian distribution with a mean value, $\mu = 0$, and standard deviation, $\sigma = 1$.

Occasionally published papers may be found that use expressions such as *standard error* or *probable error*. If definitions do not accompany such expressions then they should be treated with caution, since different meanings may be attributed by different authors.

1.4 A brief history of the standardization of units in general

The history of the development of measurement units is well covered in many excellent books that range from those for children, such as Peter Patilla's *Measuring Up Size* (2000) and the lighthearted approach of Warwick Cairns in *About the Size of It* (2007), to the scholarly and comprehensive *Encyclopaedia of Scientific Units, Weights and Measures* by François Cardarelli (2003), and Ken Alder's detailed account of the original determination of the metre in the late eighteenth century, *The Measure of All Things* (2004).

Everyday units in common use from earliest times included lengths based on human anatomy, such as the length of a man's foot, the width of a hand, the width of a thumb, the length of a leg from the ground to the hip joint, and the full extent of the outstretched arms. Greater distances could be estimated by, e.g., noting the number of paces taken in walking from town A to town B. Crude standard weights were provided by a grain of barley, a stone and a handful of fruit. Early measures of dry and liquid capacity used natural objects as containers, such as gourds, large bird eggs and sea shells. Given that many such units were either qualitative or dependent on whose body was being used (e.g., King Henry I of England decreed in 1120 that the yard should be the distance from the tip of his nose to the end of his outstretched arm), trading from one village to another could be fraught with difficulties and even lead to violent altercations.

One of the earliest records of attempted standardization to assist in trade is set out in the Hebrew Bible in Leviticus, 19, 35–36 (Moffatt, 1950): 'You must never act dishonestly, in court, or in commerce, as you use measures of length, weight, or capacity; you must have accurate balances, accurate weights, and an honest measure for bushels and gallons.'

Around 2000 years later, King John of England and his noblemen inserted a clause in the Magna Carta (number 35 on the Salisbury Cathedral copy of the document) that stated (in translation from the original abbreviated Latin text): 'Let there be throughout our kingdom a single measure for wine and for ale and for corn, namely: the London quarter,[4] and a single width of cloth (whether dyed, russet or halberjet)[5] namely two ells within the selvedges; and let it be the same with weights as with measures.'

It would appear to be very difficult to introduce a new set of standard units by legislation. Even the French, under Emperor Napoleon I, preferred a mainly non-decimal system, which had more than 250 000 different weights and measures with 800 different names, to the elegant simplicity of the decimal metric system. This preference caused Napoleon to repeal the act governing the use of the metric system (passed by the republican French National Assembly in 1795, instituting the Système Métrique Décimal) and allowing the return to the *ancien régime* in 1812. The metric system finally won out in 1837 when use of the units was made compulsory. One hundred and sixty years later it was the turn of the British to object to the introduction of the metric system, despite such a change greatly simplifying calculations using both distance and weight measurements.

1.5 A brief history of the standardization of scientific units

With the beginnings of modern scientific measurements in the seventeenth century, the scientists of the time began to appreciate the value and need for a standardized set of well-defined measurement units.

The first step towards a non-anthropocentric measurement system was proposed by the Abbé Gabriel Mouton, who in 1670 put forward the idea of a unit of length (which he named the *milliare*) equal to one thousandth of a minute of arc along the North–South meridian line. Mouton may fairly be considered the originator of the metric system, in that he also proposed three multiple and three submultiple units based on the milliare but differing by factors of ten, which were named by

[4] The London quarter was a measure that King Edward I of England decreed, in 1296, to be exactly eight striked bushels, where 'striked' implied the measuring container was full to the brim.

[5] 'halberjet' is an obsolete term for a type of cloth (Funk *et al.*, 1946).

adding prefixes to milliare. The premature death of Mouton prevented him from developing his work further.

The English architect and mathematician Sir Christopher Wren proposed in 1667 using the length of the seconds pendulum as a fixed standard, an idea that was supported by the French astronomer Abbé Jean Picard in 1671, the Dutch astronomer Christiaan Huygens in 1673 and the French geodesist Charles Marie de la Condamine in 1746. Neither the milliare nor the length of the seconds pendulum was chosen to be the standard of length however, with that honour going to a measurement of length based on a particular fraction of the circumference of the Earth.

The metre, as the new unit of length was named, was originally defined as one ten millionth part of the distance from the North Pole to the Equator along a line that ran from Dunkirk through Paris to Barcelona. The survey of this line was carried out under the direction of P. F. E. Méchain and J. B. J. Delambre. Both were astronomers by profession, who took from 1792 to 1799 to complete the task. A comparison with modern satellite measurements produces a difference of 0.02%, with the original determined metre being 0.2 mm too short (Alder, 2004). A platinum rod was made equal in length to the metre determined from the survey and deposited in the Archives de la République in Paris. It was accompanied by a one-kilogram mass of platinum, as the standard unit of mass, in the first step towards the establishment of the present set of SI units. Following the establishment of the Conférence Général des Poids et Mesures (CGPM) in 1875, construction of new platinum–iridium alloy standards for the metre and kilogram were begun.

The definition of the metre based on the 1889 international prototype was replaced in 1960 by one based upon the wavelength of krypton 86 radiation, which in turn was replaced in 1983 by the current definition based on the length of the path travelled by light, in vacuo, during a time interval of 1/(299 792 458) of a second.

The unit of time, the second, initially defined as 1/(86 400) of a mean solar day was refined in 1956 to be 1/(31 556 925.974 7) of the tropical year for 1900 January 0 at 12 h ET (Ephemeris Time). This astronomical definition was superseded by 1968, when the SI second was specified in terms of the duration of 9 192 631 770 periods of the radiation corresponding to the transition between two hyperfine levels of the ground state of the caesium 133 atom at a thermodynamic temperature of 0 K.

The unit of mass (kilogram) is the only SI unit still defined in terms of a manufactured article, in this case the international prototype of the kilogram which, with the metre, were sanctioned by the first Conférence Général des Poids et Mesures (CGPM) in 1889. These joined the astronomically determined second to form the basis of the mks (metre–kilogram–second) system, which was similar to the cgs (centimetre–gram–second) system proposed in 1874 by the British Association for the Advancement of Science.

A move to incorporate the measurement of other physical phenomena into the metric system was begun by Gauss with absolute measurements of the Earth's magnetic field using the millimetre, gram and second. Later, collaborating with Weber, Gauss extended these measures to include the study of electricity. Their work was further extended in the 1860s by Maxwell and Thomson and others working through the British Association for the Advancement of Science (BAAS). Ideas that were incorporated at this time include the use of unit-name prefixes from micro to mega to signify decimal submultiples or multiples.

In the fields of electricity and magnetism, the base units in the cgs system proved too small and, in the 1880s, the BAAS and the International Electrotechnical Commission produced a set of *practical units*, which include the ohm (resistance), the ampere (electric current) and the volt (electromotive force). The cgs system for electricity and magnetism eventually evolved into three subsystems: esu (electrostatic), emu (electromagnetic) and practical. This separation introduced unwanted complications, with the need to convert from one subset of units to another.

This difficulty was overcome in 1901 when Giorgi combined the mechanical units of the mks system with the practical electrical and magnetic units. Discussions at the 6th CGPM in 1921 and the 7th CGPM in 1927 with other interested international organizations led, in 1939, to the proposal of a four-unit system based on the metre, kilogram, second and ampere, which was approved by the CIPM in 1946.

Eight years later at the 10th CGPM, the ampere (electric current), kelvin (thermodynamic temperature) and candela (luminous intensity) were introduced as base units. The full set of six units was named the *Système International d'Unités* by the 11th CGPM in 1960. The final base unit in the current set, the mole (amount of substance), was added at the 14th CGPM in 1971.

The task of ensuring the worldwide unification of physical measurements was given to the Bureau International des Poids et Mesures (BIPM) when it was established by the 1875 meeting of the Convention du Mètre. Seventeen states signed the original establishment document. The functions of the BIPM are as follows:

1. Establish fundamental standards and scales for the measurement of the principal physical quantities and maintain the international prototypes.
2. Carry out comparisons of national and international prototypes.
3. Ensure the coordination of corresponding measuring techniques.
4. Carry out and coordinate measurements of the fundamental physical constants relevant to these activities.

This brief enables the BIPM to make recommendations to the appropriate committees concerning any revisions of unit definitions that may be necessary.

1.6 The future of SI units

It is evident from reading the above section on the development of units of measurement that it is very much an ongoing project. Today's definitions and measurements are generally the best that are available now, but that is no guarantee that they will still be so tomorrow.

The most recent candidates for change are the kilogram, the ampere, the kelvin, and the mole. Under discussion are changing the definition of these units in the following ways:

1. The kilogram is a unit of mass such that the Planck constant is exactly $6.626\,069\,3 \times 10^{-34}$ J . s (joule seconds).
2. The ampere is a unit of electric current such that the elementary charge is exactly $1.602\,176\,53 \times 10^{-19}$ coulombs (where 1 coulomb = 1 A . s (ampere second)).
3. The kelvin is a unit of thermodynamic temperature such that the Boltzmann constant is exactly $1.380\,650\,5 \times 10^{-23}$ J . K^{-1} (joules per kelvin).
4. The mole is an amount of substance such that the Avogadro constant is exactly $6.022\,141\,5 \times 10^{23}$ mol^{-1}.

No changes are under immediate consideration for the standard units of time (second), length (metre) or luminous intensity (candela).

1.7 Summary and recommendations

1.7.1 Summary

Astronomers tend to use a variety of units when describing the same quantity, which has the potential to lead to confusion and mistakes. To a certain extent this reflects the different training they may have received and the subsequent areas of astronomy in which they have carried out research. The advantages of standard systems in everyday use, as well as in astronomy, has unfortunately not meant that such systems have been readily adopted by those they were designed to benefit.

The international system is currently the most widely accepted and used set of physical units. It is still undergoing changes to the base unit definitions, mainly to tie them to fundamental physical constants rather than Earth-based constants or prototype model representations.

1.7.2 Recommendations

For astronomers who do not routinely use SI units, a simple approach to doing so would be to convert whatever units are being used into their SI equivalents and prepare research papers, presentations and lecture notes with the final results given in both forms. This would lead to SI units becoming familiar, acceptable and, hopefully, the universal system of choice.

2

An introduction to SI units

The name Système International d'Unités (International System of Units), with the abbreviation SI, was adopted by the 11th Conférence Générale des Poids et Mesures (CGPM) in 1960.

This system includes two classes of units:

- base units
- derived units,

which together form the coherent system of SI units.

2.1 The set of SI base units

There are seven well defined base units in the SI. They are: the second, the metre, the kilogram, the candela, the kelvin, the ampere and the mole, all selected by the CGPM and regarded, by convention, to be dimensionally independent. Table 2.1 lists the base quantities and the names and symbols of the base units. The order of the base units given in the table follows that of the chapters in this book.

2.2 The set of SI derived units

Derived SI units are those that may be expressed directly by multiplying or dividing base units, e.g., density (kg . m^{-3}) or acceleration (m . s^{-2}) or electric charge (A . s). Table 2.2 lists examples of SI derived units obtained from base units.

Special names have been assigned to selected derived units that are used to prevent unwieldy combinations of base SI names occurring. Table 2.3 gives some examples of such special names, with the derived unit expressed in terms of both other SI units and of base SI units only. Note that the radian and steradian were originally termed supplementary SI derived units.

Table 2.4 lists some examples of SI derived units whose names and symbols include SI derived units with special names and symbols.

Table 2.1. *SI base units (BIPM, 2006)*

Base quantity	Name	Symbol
time	second	s
length	metre	m
mass	kilogram	kg
luminous intensity	candela	cd
thermodynamic temperature	kelvin	K
electric current	ampere	A
amount of substance	mole	mol

Table 2.2. *SI derived units (BIPM, 2006)*

Derived quantity	Name	Symbol
area	square metre	m^2
volume	cubic metre	m^3
speed, velocity	metre per second	$m\,.\,s^{-1}$
acceleration	metre per second squared	$m\,.\,s^{-2}$
wavenumber	reciprocal metre	m^{-1}
density, mass density	kilogram per cubic metre	$kg\,.\,m^{-3}$
specific volume	cubic metre per kilogram	$m^3\,.\,kg^{-1}$
current density	ampere per square metre	$A\,.\,m^{-2}$
magnetic field strength	ampere per metre	$A\,.\,m^{-1}$
concentration (of amount of substance)	mole per cubic metre	$mol\,.\,m^{-3}$
luminance	candela per square metre	$cd\,.\,m^{-2}$

2.3 Non-SI units currently accepted for use with SI units

Table 2.5 lists examples of non-SI units that have been accepted for use with the International System.

Table 2.6 gives three non-SI units that have been determined experimentally and are also accepted for use with the International System. The first two values in the table were obtained from Mohr *et al.* (2007) and that for the astronomical unit from the USNO online ephemeris.[6] Table 2.7 lists some other non-SI units that are currently accepted for use with the International System.

[6] See page K6, http://asa.usno.navy.mil/index.html

Table 2.3. *SI derived units with special names (BIPM, 2006)*

Derived quantity	Name	Symbol	Other SI	Base SI
plane angle	radian	rad		$m\,.\,m^{-1}$
solid angle	steradian	sr		$m^2\,.\,m^{-2}$
frequency	hertz	Hz		s^{-1}
force	newton	N		$m\,.\,kg\,.\,s^{-2}$
pressure, stress	pascal	Pa	$N\,.\,m^{-2}$	$m^{-1}\,.\,kg\,.\,s^{-2}$
energy, work, quantity of heat	joule	J	$N\,.\,m$	$m^2\,.\,kg\,.\,s^{-2}$
power, radiant flux	watt	W	$J\,.\,s$	$m^2\,.\,kg\,.\,s^{-3}$
electric charge, quantity of electricity	coulomb	C		$A\,.\,s$
potential difference, electromotive force	volt	V	$W\,.\,A^{-1}$	$m^2\,.\,kg\,.\,s^{-3}\,.\,A^{-1}$
capacitance	farad	F	$C\,.\,V^{-1}$	$m^{-2}\,.\,kg^{-1}\,.\,s^4\,.\,A^2$
electrical resistance	ohm	Ω	$V\,.\,A^{-1}$	$m^2\,.\,kg\,.\,s^{-3}\,.\,A^{-2}$
magnetic flux	weber	Wb	$V\,.\,s$	$m^2\,.\,kg\,.\,s^{-2}\,.\,A^{-1}$
magnetic flux density	tesla	T	$Wb\,.\,m^{-2}$	$kg\,.\,s^{-2}\,.\,A^{-1}$
inductance	henry	H	$Wb\,.\,A^{-1}$	$m^2\,.\,kg\,.\,s^{-2}\,.\,A^{-2}$
luminous flux	lumen	lm	$cd\,.\,sr$	cd
illuminance	lux	lx	$lm\,.\,m^{-2}$	$cd\,.\,m^{-2}$
activity	bequerel	Bq		s^{-1}

2.4 Other non-SI units

In astronomy, many derived cgs units are in common use. The relationships between these and their values in SI units are set out in Table 2.8.

The cgs system is based on the centimetre, the gram and the second, with other units expressed in terms of these quantities. Unfortunately, electrical and magnetic units can be expressed in three different ways leading to three different systems: the cgs electrostatic system, the cgs electromagnetic system and the cgs Gaussian system.

A final set of examples, given in Table 2.9, is of units that were more common in older texts. If still used it is essential that the unit be redefined in SI terms.

2.5 Prefixes to SI units

In science in general and astronomy and astrophysics in particular, a huge range of numbers is covered from, for example, the Planck time ($\sim 5.4 \times 10^{-44}$ s), the time before which it is currently not possible to describe phenomena that might be

Table 2.4. *Some SI derived units whose names and symbols include SI derived units with special names and symbols (BIPM, 2006)*

Derived quantity	Name	Symbol	Base SI
dynamic viscosity	pascal second	Pa . s	m^{-1} . kg . s^{-1}
moment of force	newton metre	N . m	m^2 . kg . s^{-2}
surface tension	newton per metre	N . m^{-1}	kg . s^{-2}
angular velocity	radian per second	rad . s^{-1}	s^{-1}
angular acceleration	radian per second squared	rad . s^{-2}	s^{-2}
heat flux density, irradiance	watt per square metre	W . m^{-2}	kg . s^{-3}
specific energy	joule per kilogram	J . kg^{-1}	m^2 . s^{-2}
thermal conductivity	watt per metre kelvin	W . (m . K)$^{-1}$	m . kg . s^{-3} . K^{-1}
energy density	joule per cubic metre	J . m^{-3}	m^{-1} . kg . s^{-2}
electric field strength	volt per metre	V . m^{-1}	m . kg . s^{-3} . A^{-1}
electric charge density	coulomb per cubic metre	C . m^{-3}	m^{-3} . s . A
electric flux density	coulomb per square metre	C . m^2	m^{-2} . s . A
permittivity	farad per metre	F . m^{-1}	m^{-3} . kg^{-1} . s^4 . A^2
permeability	henry per metre	H . m^{-1}	m . kg . s^{-2} . A^{-2}
molar energy	joule per mole	J . mol^{-1}	m^2 . kg . s^{-2} . mol^{-1}
radiant intensity	watt per steradian	W . sr^{-1}	m^2 . kg . s^{-3}
radiance	watt per square metre steradian	W . m^{-2} . sr^{-1}	kg . s^{-3}

Table 2.5. *Non-SI units accepted for use with the International System (BIPM, 2006)*

Name	Symbol	Value in SI units
minute	min	60 s
hour	h	3600 s
day	d	86 400 s
degree	°	$(\pi/180)$ rad
minute	′	$(\pi/10\,800)$ rad
second	″	$(\pi/648\,000)$ rad
litre	l, L	10^{-3} m^3
tonne	t	10^3 kg
neper	Np	1 Np = 1
bel	B	1 B = (1/2) ln 10 (Np)

Table 2.6. *Non-SI units accepted for use with the International System whose values in SI units are obtained experimentally (BIPM, 2006)*

Name	Symbol	Value in SI units
electronvolt	eV	$1.602\,176\,487 \times 10^{-19}$ J
unified atomic mass unit	u	$1.660\,538\,782 \times 10^{-27}$ kg
astronomical unit	au	$1.495\,978\,714\,64 \times 10^{11}$ m

Table 2.7. *Some other non-SI units currently accepted for use with the International System (BIPM, 2006)*

Name	Symbol	Value in SI units
are	a	10^2 m^2
hectare	ha	10^4 m^2
bar	bar	10^5 Pa
angstrom	Å	10^{-10} m
barn	b	10^{-28} m^2

Table 2.8. *Derived cgs units with special names (BIPM, 2006). The mathematical symbol* $\wedge$ *is used for 'corresponds to'*

Name	Symbol	Value in SI units
erg	erg	10^{-7} J
dyne	dyn	10^{-5} N
poise	P	10^{-1} Pa . s
stokes	St	10^{-4} m^2 . s^{-1}
gauss	G	$\wedge\ 10^{-4}$ T
oersted	Oe	$\wedge\ (1000/4\pi)$ A . m^{-1}
maxwell	Mx	$\wedge\ 10^{-8}$ Wb
stilb	sb	10^{-4} cd . m^{-2}
phot	ph	10^4 lx
gal	Gal	10^{-2} m . s^{-2}

Table 2.9. *Examples of other non-SI units (BIPM, 2006)*

Name	Symbol	Value in SI units
curie	Ci	3.7×10^{10} Bq
jansky	Jy	10^{-26} W . m^{-2} . Hz^{-1}
fermi		10^{-15} m
standard atmosphere	atm	101 325 Pa
micron	μ	10^{-6} m

in progress due to a lack of suitable theories), up to $\sim 10^{41}$ kg for the mass of the Milky Way galaxy. In astronomy, the range of official SI prefixes is insufficient to cover the range of numbers, which has lead to the imaginative invention of many others (not yet approved by the CGPM).

2.5.1 CGPM-approved prefixes for SI units

In order to cope with very large or very small numbers in various branches of science, engineering and everyday life, the CGPM approved the use of a set of prefixes that range from 10^{-24} up to 10^{24}. They are listed with their multiplication factors, their names and their symbols in Table 2.10.

It is evident that even this range of prefixes is insufficient to meet the needs of astronomy.

2.5.2 Unofficial prefixes for SI units

An extension to the CGPM-approved prefixes was suggested by Mayes (1994) to cover even larger numbers from 10^{27} to 10^{48} and smaller numbers from 10^{-33} to 10^{-27}. Their names, symbols and multiplying factors are given in Table 2.11. Examples of possible uses of the Mayes' prefixes are given by Atkin (2007).

An obvious difficulty with a system of 31 prefixes derived from several different languages is remembering which is which, though, due mainly to usage in computer science, the names of the larger number prefixes are becoming more familiar with increasing computer storage capacity. The following two systems offer alternative solutions.

2.5.3 Powers of 1000

Languages used (Mayes, 1994) in deriving the official SI prefixes are: Latin, Greek, Danish and Italian, with the unofficial prefixes adding: Portuguese, French, Spanish, Russian, Malay – Indonesian, Chinese, Sanskrit, Arabic, Hindi and Maori.

Table 2.10. *SI approved prefixes*

Factor	Name	Symbol	Factor	Name	Symbol
10^{-24}	yocto	y	10^{1}	deca	da
10^{-21}	zepto	z	10^{2}	hecto	h
10^{-18}	atto	a	10^{3}	kilo	k
10^{-15}	femto	f	10^{6}	mega	M
10^{-12}	pico	p	10^{9}	giga	G
10^{-9}	nano	n	10^{12}	tera	T
10^{-6}	micro	μ	10^{15}	peta	P
10^{-3}	milli	m	10^{18}	exa	E
10^{-2}	centi	c	10^{21}	zetta	Z
10^{-1}	deci	d	10^{24}	yotta	Y

Table 2.11. *Mayes unofficial prefixes*

Factor	Name	Symbol	Factor	Name	Symbol
10^{-33}	weto	w	10^{27}	nava	N or nv
10^{-30}	vindo	v	10^{30}	sansa	S or sa
10^{-27}	tiso	t	10^{33}	besa	B or be
			10^{36}	vela	V or ve
			10^{39}	astra	A or at
			10^{42}	cata	C or ca
			10^{45}	quinsa	Q or qu
			10^{48}	ultra	U or ut

Whilst being a worthy attempt to produce at least one prefix derived from a language originating from each of the six continents there is no obvious connection between the prefix and the multiplying factor it is meant to represent. A possible way around this problem is simply to use the straightforward and commonly used names (e.g., million, billion, trillion, quadrillion etc.) assigned to the numbers. For numbers smaller than 1 adding 'th' after the name (e.g., millionth, billionth, trillionth, quadrillionth etc.) suffices. Table 2.12 sets out a list of multipliers for SI units that run from 10^{-48} (one quindecillionth) to 10^{48} (one quindecillion).

Note that there is a simple relationship between the power, n, to which 1000 is raised and the Latin prefix of the ending '-illion'. In Table 2.12, each line, starting at the top-left-hand side of the table, is 1000 times smaller than the values in the next line in the table. Unfortunately, the prefixes tend to become cumbersome and

Table 2.12. *Powers of 1000*

Factor (1000^n)	Name	n	Factor (1000^n)	Name	n
10^{-48}	quindecillionth	−16	10^{3}	thousand	1
10^{-45}	quattuordecillionth	−15	10^{6}	million	2
10^{-42}	tredecillionth	−14	10^{9}	billion	3
10^{-39}	duodecillionth	−13	10^{12}	trillion	4
10^{-36}	undecillionth	−12	10^{15}	quadrillion	5
10^{-33}	decillionth	−11	10^{18}	quintillion	6
10^{-30}	nonillionth	−10	10^{21}	sextillion	7
10^{-27}	octillionth	−9	10^{24}	septillion	8
10^{-24}	septillionth	−8	10^{27}	octillion	9
10^{-21}	sextillionth	−7	10^{30}	nonillion	10
10^{-18}	quintillionth	−6	10^{33}	decillion	11
10^{-15}	quadrillionth	−5	10^{36}	undecillion	12
10^{-12}	trillionth	−4	10^{39}	duodecillion	13
10^{-9}	billionth	−3	10^{42}	tredecillion	14
10^{-6}	millionth	−2	10^{45}	quattuordecillion	15
10^{-3}	thousandth	−1	10^{48}	quindecillion	16
10^{0}	one	0			

the displacement of the value of n by one relative to the Latin prefix offers the opportunity for mistakes to be made.

2.5.4 Some astronomical examples

The values of astronomical quantities used in this section were taken from Cox (2000).

Planck time $\simeq 5.4 \times 10^{-44}$ s (second) (SI base unit)
or 5.4×10^{-20} ys (yoctosecond) (official SI prefix)
or 5.4×10^{-11} ws (wetosecond) (Mayes' unofficial prefix)
or 5.4 quattuordecillionths s (powers of 1000)
wavelength of γ - radiation $\simeq 1 \times 10^{-14}$ m (metre) (SI base unit)
or 10 fm (femtometre) (official SI prefix)
or 10 quadrillionths m (powers of 1000)
1 astronomical unit $\simeq 1.496 \times 10^{11}$ m (metre) (SI base unit)
or 149.6 Gm (gigametre) (official SI prefix)
or 149.6 billion m (powers of 1000)

1 parsec $\simeq 3.086 \times 10^{16}$ m (metre) (SI base unit)
or 30.86 Pm (petametre) (official SI prefix)
or 30.86 quadrillion m (powers of 1000)
1 solar mass $\simeq 1.989 \times 10^{30}$ kg (kilogram) (SI base unit)
or 1.989×10^{9} Yg (yottagram) (official SI prefix)[7]
or 1.989 Bg (besagram) (Mayes' unofficial prefix)
or 1.989 nonillion kg (powers of 1000)
mass of Milky Way Galaxy $\simeq 1.89 \times 10^{41}$ kg (kilogram) (SI base unit)
or 1.89×10^{20} Yg (yottagram) (official SI prefix)
or 189 Cg (catagram) (Mayes' unofficial prefix)
or 189 duodecillion kg (powers of 1000)
luminous intensity of an $M_V = 0$ star outside the Earth's atmosphere $\simeq 2.45 \times 10^{29}$ cd (candela) (SI base unit)
or 2.45×10^{5} Ycd (yottacandela) (official SI prefix)
or 245 Ncd (navacandela) (Mayes' unofficial prefix)
or 245 octillion cd (powers of 1000)

2.5.5 *Other methods of denoting very large or very small numbers*

It is evident that none of the proposed modifiers to the name of the unit is satisfactory for astronomers. A simpler way of writing and speaking powers of ten would appear to be the answer, e.g., instead of writing 2×10^{30} kg the expression 2 d 30 kg could be used, where 'd', standing for deca is both the Greek word $\delta\epsilon\kappa\alpha$ for ten (Liddell & Scott, 1996), and the SI prefix for 10, which would assume the meaning '10 to the power'. So, in the example given, instead of saying 'two times ten to the power thirty', the shorter 'two d thirty' would be used. Examples of the **d** notation are given in many of the tables throughout this book.

Allen (1951) proposed the use of 'dex' for the logarithm to base 10 of a number so that 2×10^{30} would be written as 30.301 dex.

Urry (1988), tackling problems associated with objects of galactic mass, used the compact shorthand M_8 instead of $10^8\,M_\odot$ with M_n representing a mass of 10^n solar masses.

2.6 IAU recommendations regarding SI units

The International Astronomical Union in its style manual (Wilkins, 1989)[8] states that: 'The international system (SI) of units, prefixes and symbols should be used for all physical quantities except that certain special units, may be used in astronomy,

[7] remember that 1000 g = 1 kg.
[8] See also www.iau.org/science/publications/proceedings_rules/units/

Table 2.13. *Non-SI units recognized for use in astronomy*

Name	Symbol	Value in SI units
Julian year	a	$3.155\,76 \times 10^{7}$ s
cycle	c	$2\,\pi$ rad
astronomical unit	au	149.598×10^{9} m
parsec	pc	30.857×10^{15} m
solar mass	$M_{\odot}$	1.9891×10^{30} kg
atomic mass unit	u	$1.660\,539 \times 10^{-27}$ kg

Table 2.14. *Obsolete units that should not be used*

Name	Symbol	Value in SI units
angstrom	Å	10^{-10} m
micron	μ	10^{-6} m
fermi		10^{-15} m
barn	b	10^{-28} m^{2}
cubic centimetre	cc	10^{-6} m^{3}
dyne	dyn	10^{-5} N
erg	erg	10^{-7} J
calorie	cal	4.1868 J
bar	bar	10^{5} Pa
standard atmosphere	atm	101.325 kPa
gal	Gal	10^{-2} m . s^{-2}
eotvos	E	10^{-9} s^{-2}
gauss	G	10^{-4} T
gamma		10^{-9} T
oersted	Oe	$(1000/4\pi)$ A . m^{-1}

without risk of confusion or ambiguity, in order to provide a better representation of the phenomena concerned.'

In addition to the non-SI units listed in Table 2.5 and Table 2.6, the non-SI units given in Table 2.13 are recognized for use in astronomy:

The IAU recommendations for prefixes are in line with those of BIPM and are as given in Table 2.10.

Non-SI units such as British Imperial, American or other national systems of units should not be used. The cgs and obsolete units given in Table 2.14 should not be used, though some are still currently accepted for use with the International System (BIPM, 2006).

2.6.1 Angle

Currently a sexagesimal-based system of units is used to specify the positions of celestial bodies in astronomy. Declination values are typically given as degrees, minutes and seconds of arc north or south of the celestial equator, and right ascension as hours, minutes and seconds of time increasing eastwards from zero at the intersection point of the celestial equator and the ecliptic, marking the position of the Sun at the Vernal Equinox (where the declination of the Sun moves from negative values south of the equator to positive north of the equator). This use of a unit of time to represent an angle is a possible source of confusion and should be avoided. Detailed relationships between the SI unit of angle (the radian) and the various sexagesimal systems are given in Chapter 4. If, for some reason, the radian is not considered a suitable unit in a particular circumstance, then the degree with decimal subdivision should be used.

The use of the 'mas' meaning milliarcsecond for angular resolution or angular separation of astronomical objects should be replaced by a more appropriate SI unit, such as the nrad (nanoradian).

2.6.2 Time

Other than the base SI unit of time, the second, astronomers use longer lengths of time, such as the minute (60 seconds), the hour (3600 seconds), the day (86 400 seconds) and the Julian year consisting of 365.25 d or $31.557\,6 \times 10^6$ s. There are, however, several different kinds of day and year that relate to particular problems in astronomy (see Chapter 5).

The variability of the Earth's rotation rate means that time based on that rate varies with respect to the SI second. Hence sidereal, solar and Universal Time should be considered as measurements of hour angle expressed in time measure and not suitable for precise measures of time intervals.

2.6.3 Distance and mass

The IAU accepts the use of a special set of length, mass and time units for the study of motions in the Solar System – they are related to one another through the adopted value of the Gaussian gravitational constant k ($=0.017\,202\,098\,95$) when it is expressed in these units.

The astronomical unit (the unit of distance) is the radius of a circular orbit in which a body of negligible mass, and free from perturbations, would revolve around the Sun in $2\pi/k$ days. This distance is slightly less than the semimajor axis of the Earth's orbit ($\sim 1.495\,978 \times 10^{11}$ m).

The parsec is a distance ($\sim 3.086 \times 10^{16}$ m) equal to that at which the astronomical unit subtends an angle of 1 arcsec ($\pi/(180 \times 3600)$ rad).

The light year is the distance ($\sim 9.461 \times 10^{15}$ m) travelled by light, in vacuo, in one Julian year.

The astronomical unit of mass is the solar mass ($\sim 1.989\,1 \times 10^{30}$ kg) denoted by $M_\odot$.

2.6.4 *Wavenumber*

The reciprocal wavelength or wavenumber is used mainly by infrared astronomers and is normally based on the cm^{-1}. If used it should be in the SI unit form of m^{-1}, but in either case the unit must be given as it is not dimensionless.

2.6.5 *Magnitude*

Magnitude may be defined as the ratio of the logarithm of the signal strength of the celestial object of interest to that of a standard star or object. As such it is a dimensionless quantity. Some magnitude scales have been calibrated in terms of SI units (see Chapter 8).

2.7 Summary and recommendations

2.7.1 *Summary*

The Système International d'Unités consists of two classes of units: the base units of time, length, mass, luminous intensity, thermodynamic temperature, electric current and amount of substance; and a set of derived units obtained by dividing or multiplying base units. The SI units are decimal, with larger and smaller multiples of the base units assigned prefixes. The current set of official prefixes is insufficient to meet the needs of astronomers.

The International Astronomical Union has produced a set of recommendations concerning the use of SI units in astronomy and the continued use of specialized non-SI astronomical units.

2.7.2 *Recommendations*

Astronomers should follow the recommendations proposed by the IAU with particular reference to dropping the use of the cgs-based units so prevalent at present. Given that the current set of official SI prefixes is inadequate, consistency may be maintained by quantities being presented as the product of the measurement times the basic unit (i.e., the unit without any prefix). For example, the astronomical unit should be given as 1.496×10^{11} m or 1.496 d 11 m and not 149.6 Gm. Examples of all three forms are given in different tables throughout the book.

3

Dimensional analysis

3.1 Definition of dimensional analysis

Dimensional analysis is a technique for studying the dimensions of physical quantities. It may be used to:

1. Reduce the physical properties of derived SI units into those of the more fundamental SI base units.
2. Assist in converting quantities expressed in non-SI units to SI units.
3. Verify the correctness of an equation in terms of dimensional and unitary consistency.
4. Determine the dimension and unit of a variable in an equation.
5. Provide a means for selecting relevant data and how best to present it.

3.1.1 The dimensions of the SI base units

The seven base SI units provide seven independent dimensions with which to describe any derived SI unit. The symbols for each of the base quantity dimensions are given in Table 3.1.

3.1.2 Dimensions of some of the SI derived units

The dimension of a derived unit, X, is the product of the base unit dimensions, B_i, where $\mathrm{B}_i = T, L, M, J, \Theta, I$, and N, such that:

$$\dim(X) = \prod_{i=1}^{7} \dim^{\alpha_i}(B_i) \tag{3.1}$$

By way of example, the derived quantity force, P, has the derived SI unit $\mathrm{kg\,.\,m\,.\,s^{-2}}$

$$\dim(P) = [T]^{-2}.[L]^{1}.[M]^{1}.[J]^{0}.[\Theta]^{0}.[I]^{0}.[N]^{0} \tag{3.2}$$

Table 3.1. *Symbols for SI base quantity dimensions (BIPM, 2006)*

Base quantity	Dimension symbol
time	$[T]$
length	$[L]$
mass	$[M]$
luminous intensity	$[J]$
thermodynamic temperature	$[\Theta]$
electric current	$[I]$
amount of substance	$[N]$

which simplifies to:

$$\dim(P) = [L].[M].[T]^{-2} \tag{3.3}$$

Table 3.2 gives examples of the dimensions of some derived SI units. Note that both the radian and steradian are examples of dimensionless units, as in both cases the value of the dimension is 1. Some derived quantities have the same dimension, e.g., the hertz and the bequerel both have the dimension $[T]^{-1}$.

3.2 Dimensional equations

For an equation describing a physical situation to be true it is necessary that both sides of the equation have the same dimensions, that is, the equation must be dimensionally homogeneous, e.g.:

$$v = d/t \tag{3.4}$$

where $v =$ velocity (dimension $[L].[T]^{-1}$), $d =$ distance (dimension $[L]$) and $t =$ time (dimension $[T]$) may be written in dimensional terms as:

$$\dim(v) = [L].[T]^{-1} = [L]/[T] = \dim(d/t) \tag{3.5}$$

so the equation is homogeneous. Note that the requirement is for the dimensions to be consistent, so that any set of consistent units within a particular dimension may be used and converted to any other by means of a constant factor. If physical quantities have the same dimensions, they may only be combined by addition or subtraction. For example, $[L] + [L]$ or $[L] - [L]$ but not $[L] \times [L]$ (which is a measure of area), nor $[L]/[L]$ (which is a ratio and a dimensionless number).

Table 3.2. *Dimensions of some derived SI units*

Derived quantity	Name	Dimension
plane angle	radian	$[L].[L]^{-1}$
solid angle	steradian	$[L]^2.[L]^{-2}$
frequency	hertz	$[T]^{-1}$
force	newton	$[L].[M].[T]^{-2}$
pressure, stress	pascal	$[L]^{-1}.[M].[T]^{-2}$
energy, work, quantity of heat	joule	$[L]^2.[M].[T]^{-2}$
power, radiant flux	watt	$[L]^2.[M].[T]^{-3}$
electric charge, quantity of electricity	coulomb	$[T].[I]$
potential difference, electromotive force	volt	$[L]^2.[M].[T]^{-3}.[I]^{-1}$
capacitance	farad	$[L]^{-2}.[M]^{-1}.[T]^4.[I]^2$
electrical resistance	ohm	$[L]^2.[M].[T]^{-3}.[I]^{-2}$
magnetic flux	weber	$[L]^2.[M].[T]^{-2}.[I]^{-1}$
magnetic flux density	tesla	$[M].[T]^{-2}.[I]^{-1}$
inductance	henry	$[L]^2.[M].[T]^{-2}.[I]^{-2}$
luminous flux	lumen	$[J]$
illuminance	lux	$[L]^{-2}.[J]$
activity	bequerel	$[T]^{-1}$
area	square metre	$[L]^2$
volume	cubic metre	$[L]^3$
speed, velocity	metre per second	$[L].[T]^{-1}$
acceleration	metre per second squared	$[L].[T]^{-2}$
wavenumber	reciprocal metre	$[L]^{-1}$
density, mass density	kilogram per cubic metre	$[M].[L]^{-3}$
specific volume	cubic metre per kilogram	$[L]^3.[M]^{-1}$
current density	ampere per square metre	$[I].[L]^{-2}$
magnetic field strength	ampere per metre	$[I].[L]^{-1}$
concentration (of amount of substance)	mole per cubic metre	$[N].[L]^{-3}$
luminance	candela per square metre	$[J].[L]^{-2}$
rate of cooling	kelvin per second	$[\Theta].[T]^{-1}$

A dimensional equation that is to be used for converting one set of units within a dimension to another (e.g., inches to metres) must also include a conversion factor, k (to convert inches to metres, $k = 0.0254$). So Equation (3.1) above would be rewritten in the more general form:

$$\dim(X) = \prod_{i=1}^{7} [k_i \dim(B_i)]^{\alpha_i} \tag{3.6}$$

where the k_i relate to the conversion factors required to convert the units in which the measurements were carried out to base units (e.g., the multiplicative conversion factor from inches to metres (Kaye & Laby, 1959) is 1/39.370 147 = 0.025 399 96).

Worked example: convert quantities expressed in non-SI units to SI units

The orbital speed of the Earth about the Sun was commonly given as 66 600 mph (miles per hour) in older English-language popular astronomy books (e.g., Evans, 1954). What does this speed equate to in the SI base units metres and seconds?

Using Equation (3.4) above, the following dimensional equation may be written where the left-hand side is the dimension of speed:

$$\dim(v) = [k_L \,.\, L]^1 \,.\, [k_T \,.\, T]^{-1} \tag{3.7}$$

k_L is the length conversion factor from miles to metres and k_T is the time conversion factor from hours to seconds.

1 mile = 1609.344 m and 1 hour = 3600 s, so

$$\dim(v) = [1609.344\, L]^1.[3600\, T]^{-1} = 0.447\,04\,[L]^1.[T]^{-1} \tag{3.8}$$

The combined conversion factor, 0.447 04, is then multiplied by the Earth's orbital speed in mph (66 600) to give the speed in $\mathrm{m\,.\,s^{-1}}$.

Hence, 66 600 mph $\equiv 2.977 \times 10^4\ \mathrm{m\,.\,s^{-1}} \equiv 29.77\ \mathrm{km\,.\,s^{-1}}$.

Example: compare the consistency between two units of monochromatic flux density

The two units to be compared are the jansky ($10^{-26}\,\mathrm{W\,.\,m^{-2}\,.\,Hz^{-1}}$) and the ($\mathrm{erg\,.\,cm^{-2}\,.\,s^{-1}\,.\,Å^{-1}}$) used in a *Catalogue of Stellar Ultraviolet Fluxes* (Thompson *et al.*, 1978).

The component parts of the jansky are the watt, W, which is a derived unit of power equal to $1\,\mathrm{J\,.\,s^{-1}}$ (J = joule); the joule is a derived unit of energy or work and is equal to $1\,\mathrm{m^2\,.\,kg\,.\,s^{-2}}$; the hertz is a unit of frequency and is measured in inverse seconds, $\mathrm{s^{-1}}$. Combining these components, the jansky in SI base units is: $(\mathrm{m^2\,.\,kg\,.\,s^{-2}})\,.\,\mathrm{s^{-1}\,.\,m^{-2}}\,.\,(\mathrm{s^{-1}})^{-1}$, which reduces to $\mathrm{kg\,.\,s^{-2}}$, or in dimensional terms:

$$\dim(f_{\nu(Jy)}) = [M]^1 \,.\, [T]^{-2} \tag{3.9}$$

The ultraviolet unit of Thompson *et al.* (1978), hereafter called the 'TD1' unit, is a cgs unit with the following named components. The erg is the cgs unit of energy equal to $1\,\mathrm{cm^2\,.\,g\,.\,s^{-2}}$ or 10^{-7} J and the angstrom is a unit of length equal to 10^{-10} m. Combining all the components, the TD1 unit in cgs base units is $(\mathrm{cm^2\,.\,g\,.\,s^{-2}})\,.\,\mathrm{cm^{-2}\,.\,s^{-1}\,.\,cm^{-1}}$, which reduces to $\mathrm{g\,.\,s^{-3}\,.\,cm^{-1}}$ or, in dimensional terms:

$$\dim(f_{\lambda(\mathrm{TD1})}) = [M]^1 \,.\, [T]^{-3}.[L]^{-1} \tag{3.10}$$

Quite evidently, even the application of multiplicative scaling factors would not make the two monochromatic flux units compatible in a dimensional sense. The difference in derived dimensions in this case is due to the different variable used to specify the bandwidth of the observation. For the jansky, the bandwidth is measured in hertz ($[T]^{-1}$) and for the TD1 unit the bandwidth is measured in angstroms ($[L]$). To compare the measures directly, the wavelength bandwidth in angstroms needs to be converted to a frequency bandwidth in hertz. The algebraic relationship between frequency (ν) and wavelength (λ) is given by:

$$\lambda = \frac{c}{\nu} \tag{3.11}$$

where c is the velocity of light. The relationship between the bandwidths in wavelength form ($\Delta\,\lambda$) and frequency form ($\Delta\,\nu$) is obtained by differentiating Equation (3.11) to give:

$$\Delta\,\lambda = -\frac{c}{\nu^2}\,\Delta\,\nu \tag{3.12}$$

The dimensional form of (c/ν^2) is $([L].[T]^{-1}/[T]^{-2})$ or $[L].[T]$, so multiply $\dim(f_{\lambda(\mathrm{TD1})})$ by $[L].[T]$ to obtain:

$$\dim(f_{\lambda(\mathrm{TD1})}).[L]^1.[T]^1 = [M]^1.[T]^{-3}.[L]^{-1}.[L]^1.[T]^1 \tag{3.13}$$

$$= [M]^1.[T]^{-2} \tag{3.14}$$

$$= \dim(f_{\nu(\mathrm{Jy})}) \tag{3.15}$$

The negative sign in Equation (3.12) is cancelled because the value for the frequency decreases with increasing wavelength, so if $\Delta\,\lambda$ is positive then $\Delta\,\nu$ must be negative. Now all that has to be done is evaluate the multiplying factor to convert the TD1 unit to janskys.

Example: how to convert TD1 cgs flux units to janskys (derived SI unit) for the B5V star HD74071

The ESRO satellite TD1 measured ultraviolet flux of HD74071 is $f_\lambda = 24.73 \times 10^{-11}\,\mathrm{erg.s^{-1}.cm^{-2}.\AA^{-1}}$ at a wavelength of $\lambda = 1565\,\mathrm{\AA}$ with a bandwidth of $\Delta\lambda = 330\,\mathrm{\AA}$. The frequency ν, corresponding to the wavelength 1565Å derived by using equation (3.11) is 1.916×10^{15} Hz, with a frequency bandwidth derived from Equation (3.12) of 4.043×10^{14} Hz.

The TD1 flux unit may be converted to SI base units, with appropriate factors, as:

$$\begin{aligned}\text{TD1 flux unit} &= \mathrm{erg.s^{-1}.cm^{-2}.\AA^{-1}}\\ &\equiv (10^{-7})\,\mathrm{W}.(10^{-2})^{-2}.\mathrm{m^{-2}}.(10^{-10})^{-1}.\mathrm{m^{-1}}\\ &= 10^{7}.\mathrm{W.m^{-2}.m^{-1}}\end{aligned} \tag{3.16}$$

Note that the powers to which the conversion factors are raised match those of the SI units. These individual conversion factors are multiplied together to form a single conversion factor for the entire derived SI unit. Now convert the bandpass from wavelength, $\Delta\lambda$, to frequency, $\Delta\nu$, to give the TD1 units in janskys:

$$f_{\nu(\mathrm{TD1})} = f_{\lambda(\mathrm{TD1})} \cdot \frac{\Delta\lambda}{\Delta\nu} \,.\, 10^{26} \tag{3.17}$$

$$= 10^{7} \,.\, (24.73 \times 10^{-11}) \,.\, \frac{2.998 \times 10^{8}}{(1.916 \times 10^{15})^{2}} \,.\, 10^{26}$$

$$= 20.2\,\mathrm{Jy}$$

So the measured monochromatic flux density from the star HD74071 may be given as either 24.73×10^{-11} erg . cm^{-2} s^{-1} . Å^{-1} or 20.20 Jy.

Example: how to transform the catalogue measurement error in TD1 catalogue units to janskys

Thompson *et al.* (1978) give the measurement error in their ultraviolet fluxes in wavelength space as $\delta\, f_{\lambda\,(\mathrm{TD1})}$. To transform this value to frequency space, $\delta\, f_{\nu\,(\mathrm{TD1})}$, Equation (3.17) is rewritten as:

$$\delta\, f_{\nu\,(\mathrm{TD1})} = \delta\, f_{\lambda\,(\mathrm{TD1})} \cdot \frac{\Delta\,\lambda}{\Delta\,\nu} \,.\, 10^{26} \tag{3.18}$$

Substitute values for the variables from above, including the combined conversion factor of 10^{7}, in Equation (3.16) to give:

$$\delta\, f_{\nu(\mathrm{TD1})} = \pm 0.005\,\mathrm{Jy}$$

3.3 Summary and recommendations

3.3.1 Summary

Using the techniques of dimensional analysis to transform commonly used, but generally non-SI derived units, to SI derived or base units can be of considerable assistance in the reduction in the number of mistakes, which unfortunately are not uncommon in such an exercise. Some examples are given of such transformations in simple dynamics and ultraviolet stellar photometry. Many more worked examples are given in the chapters relating to each of the SI base units.

3.3.2 Recommendations

No matter how much care is taken in the transformation of units, mistakes do occur. Particular problems are associated with the powers to which the base units are raised and the signs and values of powers of ten. So always check any unit transformation that has been carried out and when satisfied that it is correct, check it again!

4
Unit of angular measure (radian)

4.1 SI definition of the radian

The **radian** is the angle subtended at the centre of a circle by an arc along the circumference whose length is equal to that of the radius of the circle.

The dimension of angular measure is $[L].[L]^{-1} = [1]$, its unit is the **radian** and its symbol is **rad**.

In Figure 4.1, the angle AOB is equal to one radian if $\frac{\overrightarrow{AB}}{\overrightarrow{OA}} = 1$, where OA = OB is the radius of the circle and $\overrightarrow{AB}$ is the distance along the circumference of the circle from A to B. Given that the circumference of a circle of radius r is $2\pi r$, then the number of radians in the circle is simply $2\pi r/r$ or 2π.

4.2 Commonly used non-SI units of angular measure

4.2.1 Converting from (° ′ ″) and (h m s) to radians

Positions of astronomical objects are regularly given as angles from a reference point or plane in degrees (°), minutes (′) and seconds (″), or sometimes as a measure of time from a reference point in hours (h), minutes (m) and seconds (s). There are 60″ in 1′ and 60 s in 1 m, 60′ in 1° and 60 m in 1 h and 360° or 24 h in a circle. This sexagesimal system was first used by the Babylonians (Pannekoek, 1961) more than 2000 years ago and most astronomical catalogues still use it, though some catalogues do include positions in radians, e.g., the SuperCOSMOS Sky Survey.[9]

There are 2π rad in 360°, which gives the following conversion factors from sexagesimal measures to radians:

$$1^\circ = \left(\frac{2\pi}{360}\right) = 0.017\,453\,292\,520\,\text{rad}$$

[9] See www-wfau.roe.ac.uk/sss

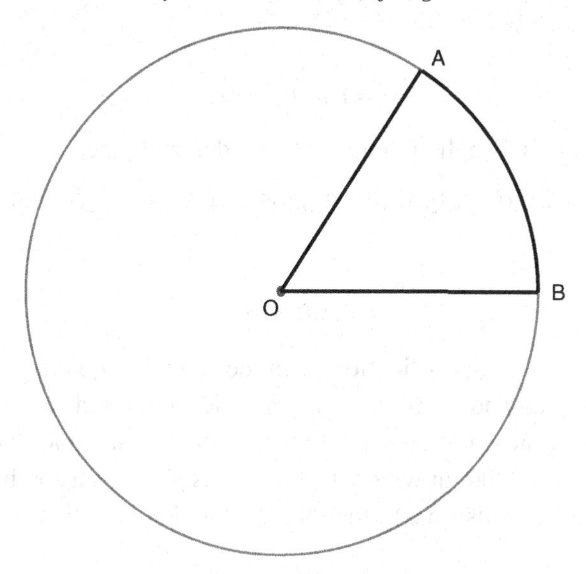

Figure 4.1. Definition of a radian.

$$1' = \left(\frac{2\pi}{60\times 360}\right) = 0.000\,290\,888\,209\,\mathrm{rad}$$

$$1'' = \left(\frac{2\pi}{60\times 60\times 360}\right) = 0.000\,004\,848\,137\,\mathrm{rad} \tag{4.1}$$

Similarly, there are 2π rad in 24 h or:

$$1\,\mathrm{h} = \left(\frac{2\pi}{24}\right) = 0.261\,799\,387\,799\,149\,\mathrm{rad}$$

$$1\,\mathrm{m} = \left(\frac{2\pi}{60\times 24}\right) = 0.004\,363\,323\,129\,985\,\mathrm{rad}$$

$$1\,\mathrm{s} = \left(\frac{2\pi}{60\times 60\times 24}\right) = 0.000\,072\,722\,052\,166\,\mathrm{rad} \tag{4.2}$$

Example: calculate the position of the bright northern star Capella in radians
Capella has the following coordinates in the FK5 catalogue for equinox (J2000.0) and epoch (J2000.0):

$$\alpha = 05^{\mathrm{h}}16^{\mathrm{m}}41.36^{\mathrm{s}} \;\; \delta = +45^{\circ}59'52.8''$$

where α is the right ascension and δ is the declination of Capella (see Section 4.3.2 below). The right ascension of Capella in radians to six decimal places is:

$$5\times 0.261\,799 + 16\times 0.004\,363 + 52.8\times 0.000\,073$$

so

$$\alpha = 1.381818\,\text{rad}$$

and the declination of Capella in radians to six decimal places is:

$$45 \times 0.017\,453 + 59 \times 0.000\,290 + 52.8 \times 0.000\,005$$

so

$$\delta = 0.802\,816\,\text{rad}$$

If, instead of measuring declination from the celestial equator (CE) northwards from 0 rad to $+\frac{\pi}{2}$ rad and southwards from 0 rad to $-\frac{\pi}{2}$ rad, we measure it from the south celestial pole (SCP) as 0 rad to the north celestial pole (NCP) as $+\pi$ rad (see Zacharias *et al.*, 2000, in which declination is given as a south polar distance in units of milliarcseconds (mas)), then the declination in radians becomes:

$$\delta_{(\text{SCP}_0)} = \delta_{(\text{CE}_0)} + \frac{\pi}{2} \tag{4.3}$$

Using the south celestial pole zero point (SCP_0), the revised declination of Capella becomes 2.373 612 rad. Expressed in milliradians (mrad), the coordinates of Capella become:

$$\alpha = 1381.817\,893\ \text{mrad}$$

$$\delta_{\text{SCP}_0} = 2373.612\,876\,\text{mrad}$$

Given that, e.g., *The Astronomical Almanac* lists its bright star positions to the nearest tenth of a second of time and nearest whole second of arc, what number of decimal places would yield a similar precision using milliradians? The answers are, approximately, three decimal places for right ascension and two for declination. So for the worked example using Capella above, a catalogue position in milliradians with similar precision to that given in *The Astronomical Almanac* would be:

$$\alpha = 1381.818\,\text{mrad}$$

$$\delta_{\text{SCP}_0} = 2373.61\,\text{mrad}$$

By way of another example, the equatorial coordinates listed in the table of 86 bright stars given in the *Handbook of the British Astronomical Association* for 2008 were converted to milliradians and the V magitudes to janskys and ln(janskys). The following equation was used to convert from magnitudes to monochromatic flux densities[10] in janskys ($10^{-26}\,\text{W}\text{m}^{-2}\,\text{Hz}^{-1}$):

$$\text{VJ} = 3600(10^{-0.2\text{V}}) \tag{4.4}$$

[10] See Chapter 8 for more details.

Table 4.1. *Equatorial coordinates in (h,m,s), (°,′,″), and milliradians, V magnitudes and monochromatic flux densities in janskys for 86 bright stars*

Star name	α h	α m	α s	δ °	δ ′	δ ″	V mag	α mrad	δ_{SCP} mrad	f_{V_ν} Jy	$\ln(f_{V_\nu})$ Jy
α And	00	08	49.7	29	08	14	2.06	38.520	2079.33	1394	7.24
β Cas	00	09	38.2	59	11	48	2.27	42.047	2603.97	1265	7.14
α Cas	00	40	59.7	56	35	02	2.23	178.874	2558.37	1289	7.16
β Cet	00	44	00.9	−17	56	24	2.04	192.051	1257.68	1407	7.24
β And	01	10	12.6	35	39	55	2.06	306.348	2193.27	1394	7.24
α Eri	01	38	01.8	−57	11	37	0.46	427.736	572.57	2912	7.97
γ And	02	04	25.5	42	22	13	2.26	542.906	2310.29	1271	7.14
α Ari	02	07	39.3	23	30	08	2.00	557.000	1980.98	1433	7.26
α UMi	02	41	49.8	89	18	03	2.02	706.116	3129.38	1420	7.25
β Per	03	08	43.5	40	59	16	2.10	823.468	2286.16	1368	7.22
α Per	03	24	56.0	49	53	27	1.79	894.190	2441.55	1578	7.36
η Tau	03	47	59.5	24	07	51	2.87	994.801	1991.95	960	6.86
α Tau	04	36	24.6	16	31	33	0.85	1206.066	1859.22	2433	7.79
β Ori	05	14	56.8	−08	11	33	0.12	1374.214	1427.81	3406	8.13
α Aur	05	17	19.1	46	00	21	0.08	1384.562	2373.74	3469	8.15
γ Ori	05	25	35.2	06	21	24	1.64	1420.639	1681.74	1691	7.43
β Tau	05	26	49.8	28	36	50	1.65	1426.064	2070.20	1683	7.42
δ Ori	05	32	26.5	−00	17	36	2.23	1450.550	1565.67	1289	7.16
ϵ Ori	05	36	38.7	−01	11	50	1.70	1468.890	1549.90	1645	7.40
ζ Ori	05	41	11.2	−01	56	19	1.77	1488.707	1536.96	1593	7.37
κ Ori	05	48	09.6	−09	40	02	2.06	1519.134	1402.07	1394	7.24
α Ori	05	55	37.9	07	24	29	0.40	1551.735	1700.09	2994	8.00
β Aur	06	00	09.1	44	56	51	1.90	1571.458	2355.27	1500	7.31
β CMa	06	23	04.4	−17	57	38	1.98	1671.472	1257.32	1446	7.27
α Car	06	24	08.5	−52	42	02	−0.72	1676.134	650.99	5015	8.52
γ Gem	06	38	12.2	16	23	29	1.93	1737.489	1856.88	1480	7.29
α CMa	06	45	31.3	−16	43	42	−1.46	1769.422	1278.83	7051	8.86
ϵ CMa	06	58	57.6	−28	59	03	1.50	1828.057	1064.92	1804	7.49
δ CMa	07	08	44.2	−26	24	26	1.86	1870.716	1109.90	1528	7.33
α Gem	07	35	08.5	31	52	09	1.95	1985.930	2127.01	1466	7.29
α CMi	07	39	44.8	05	12	10	0.38	2006.023	1661.60	3022	8.00
β Gem	07	45	50.1	28	00	18	1.14	2032.588	2059.57	2129	7.66
ζ Pup	08	03	53.0	−40	01	39	2.25	2111.339	872.18	1277	7.15
γ Vel	08	09	47.7	−47	21	43	1.78	2137.133	744.17	1585	7.36
ϵ Car	08	22	41.3	−59	32	14	1.86	2193.391	531.67	1528	7.33
δ Vel	08	44	56.3	−54	44	23	1.96	2290.475	615.40	1459	7.28
λ Vel	09	08	18.6	−43	28	02	2.21	2392.453	812.15	1301	7.17
β Car	09	13	17.4	−69	45	08	1.68	2414.183	353.39	1660	7.41
ι Car	09	17	19.0	−59	18	40	2.25	2431.752	535.62	1277	7.15
α Hya	09	28	00.3	−08	41	45	1.98	2478.389	1419.02	1446	7.27
α Leo	10	08	49.4	11	55	31	1.35	2656.492	1778.93	1933	7.56
γ Leo	10	20	26.4	19	47	54	1.99	2707.180	1916.34	1439	7.27

Table 4.1. (*cont.*)

Star name	α h	m	s	δ °	′	″	V mag	α mrad	δ_{SCP} mrad	f_{V_ν} Jy	$\ln(f_{V_\nu})$ Jy
β UMa	11	02	21.0	56	20	12	2.37	2890.047	2554.05	1208	7.10
α UMa	11	04	14.8	61	42	17	1.79	2898.322	2647.74	1578	7.36
β Leo	11	49	29.6	14	31	28	2.14	3095.748	1824.29	1343	7.20
α Cru	12	27	04.6	−63	08	46	1.33	3259.736	468.68	1951	7.57
γ Cru	12	31	38.5	−57	09	38	1.63	3279.655	573.15	1699	7.43
γ Cen	12	41	59.3	−49	00	22	2.17	3324.801	715.47	1325	7.18
β Cru	12	48	13.4	−59	44	06	1.25	3352.006	528.22	2024	7.61
ϵ UMa	12	54	24.1	55	54	50	1.77	3378.964	2546.67	1593	7.37
ζ UMa	13	24	16.0	54	52	52	2.27	3509.275	2528.65	1265	7.14
α Vir	13	25	38.5	−11	12	20	0.98	3515.274	1375.22	2292	7.73
ϵ Cen	13	40	25.9	−53	30	33	2.30	3579.808	636.88	1248	7.12
η UMa	13	47	52.5	49	16	16	1.86	3612.286	2430.73	1528	7.33
β Cen	14	04	25.8	−60	24	49	0.61	3684.520	516.37	2718	7.90
θ Cen	14	07	11.1	−36	24	41	2.06	3696.541	935.29	1394	7.24
α Boo	14	16	03.0	19	08	18	−0.04	3735.222	1904.82	3666	8.20
η Cen	14	36	03.0	−42	11	41	2.31	3822.489	834.35	1242	7.12
α Cen	14	40	10.9	−60	52	12	0.00	3840.517	508.41	3600	8.18
α Lup	14	42	29.9	−47	25	27	2.30	3850.625	743.08	1248	7.12
ϵ Boo	14	45	21.4	27	02	19	2.40	3863.097	2042.70	1192	7.08
β UMi	14	50	41.3	74	07	15	2.08	3886.361	2864.44	1381	7.23
α CrB	15	35	02.9	26	41	11	2.23	4079.918	2036.56	1289	7.16
δ Sco	16	00	50.2	−22	38	43	2.32	4192.440	1175.56	1236	7.12
α Sco	16	29	55.8	−26	27	01	0.96	4319.384	1109.15	2313	7.74
α TrA	16	49	34.3	−69	02	32	1.92	4405.087	365.78	1486	7.30
ϵ Sco	16	50	42.9	−34	18	29	2.29	4410.076	972.00	1254	7.13
λ Sco	17	34	11.2	−37	06	33	1.63	4599.757	923.11	1699	7.43
α Oph	17	35	19.8	−12	33	16	2.08	4604.745	1351.68	1381	7.23
θ Sco	17	37	55.8	−43	00	09	1.87	4616.090	820.26	1521	7.32
γ Dra	17	56	48.2	51	29	18	2.23	4698.440	2469.43	1289	7.16
ϵ Sgr	18	24	44.2	−34	22	48	1.85	4820.323	970.75	1535	7.33
α Lyr	18	37	13.6	38	47	31	0.03	4874.820	2247.84	3550	8.17
σ Sgr	18	55	47.5	−26	17	08	2.02	4955.826	1112.02	1420	7.25
β Cyg	19	31	03.9	27	58	41	3.08	5109.735	2059.10	871	6.77
α Aql	19	51	11.9	08	53	29	0.77	5197.583	1725.98	2525	7.83
γ Cyg	20	22	32.0	40	17	03	2.20	5334.307	2273.88	1307	7.17
α Pav	20	26	18.9	−56	42	26	1.94	5350.808	581.06	1473	7.29
α Cyg	20	41	43.3	45	18	40	1.25	5418.032	2361.62	2024	7.61
α Cep	21	18	46.9	62	37	18	2.44	5579.737	2663.75	1170	7.06
ϵ Peg	21	44	36.2	09	54	51	2.39	5692.405	1743.83	1197	7.08
α Gru	22	08	45.9	−46	55	10	1.74	5797.831	751.89	1615	7.38
β Gru	22	43	10.3	−46	50	24	2.11	5947.958	753.28	1362	7.21
α PsA	22	58	07.1	−29	34	37	1.16	6013.175	1054.58	2110	7.65
β Peg	23	04	11.3	28	07	45	2.42	6039.660	2061.74	1181	7.07
α Peg	23	05	11.1	15	15	04	2.49	6044.009	1836.97	1143	7.04

where the conversion factor of 3600 is from an absolute calibration of a zero magnitude star in the Johnson V band given by Bessell (1992), V is the listed value of the apparent V magnitude and VJ is the apparent monochromatic flux density of the star in janskys. A plot of the stars in Table 4.1 is shown in Figure 4.2 using Mollweide's projection.[11]

Obvious advantages of the SI radians angular measurement system are that only two numbers are required to specify the location of the celestial body, rather than six, and that all the declination values are positive when the declination of the South Celestial Pole is set equal to 0 mrad.

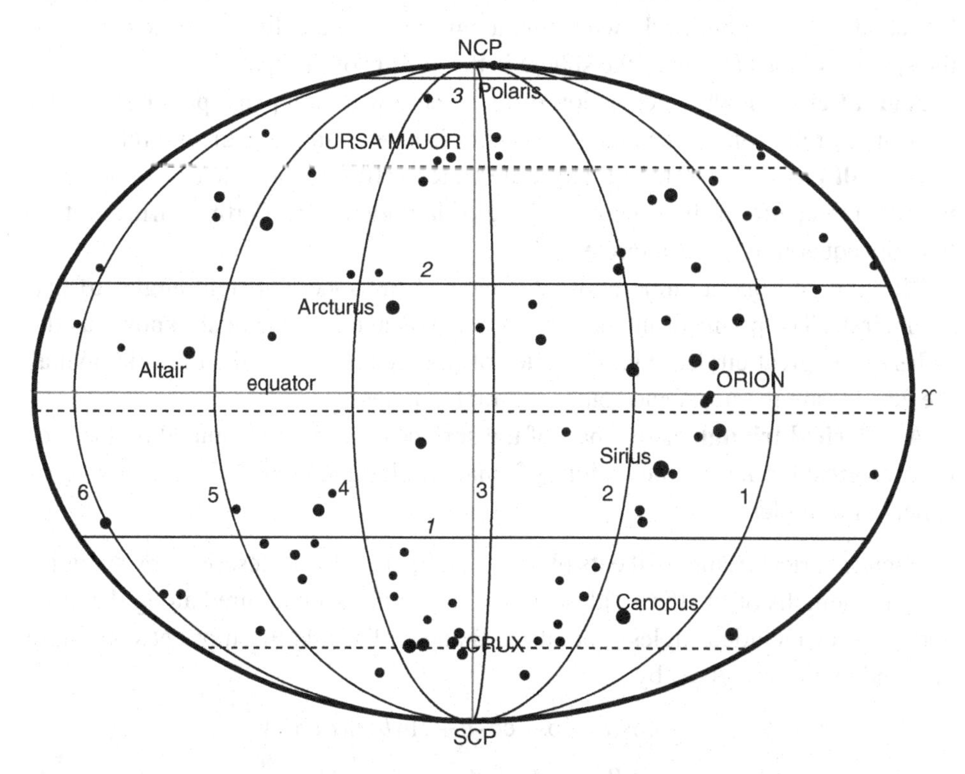

Figure 4.2. A star chart for the 86 bright stars in Table 4.1, plotted in radians, with declination measured from 0 at the south celestial pole (SCP) northwards to π rad at the north celestial pole (NCP) and right ascension increasing in an easterly direction from 0 at the First Point of Aries, ϒ, to 2π rad. Right ascensions in radians from 1 rad to 6 rad are printed to the left of the relevant circle and declinations are printed from 1 rad to 3 rad immediately adjacent to the relevant circle. The dashed declination lines are at the 0.5-rad points. The size of the star images are proportional to $\ln(f_{V_\nu})$ in janskys.

[11] See, e.g., http://www.astron.nl/aips++/docs/memos/107/node38.html

4.3 Spherical astronomy

Spherical astronomy is the study of the directions with respect to some fixed point, line or plane in which celestial bodies (e.g., planets, comets, stars, galaxies) appear to lie. As celestial bodies are, generally speaking, extremely distant, they seem to lie on the surface of a sphere, known as the **celestial sphere**.

4.3.1 Spherical triangles

Some definitions

A **sphere** is a surface formed by rotating a semicircle about its diameter through 2π radians. A straight line drawn from the mid-point of the diameter (the **centre** of the sphere) to the surface of the sphere is the **radius** of the sphere.

A **great circle** is the intersection of the sphere with any plane passing through the centre of the sphere. The Earth's equator is an example of a great circle.

A **small circle** is formed by any other intersecting surface that does not pass through the centre of the sphere. A circle of latitude on the Earth's surface, other than the equator, is a small circle.

The **axis** of a great circle is the diameter of the sphere at right angles to that great circle. The points of intersection of the axis and the sphere are known as the **poles** of the great circle. (e.g., the poles of the great circle on the Earth known as the equator are the north and south geographic poles).

A **spherical triangle** is that part of the surface of a sphere bounded by the arcs of three great circles. Combinations of small circles and great circles do not make spherical triangles.

Each spherical triangle, like its plane counterpart, may be described by six numbers: the lengths of the three sides (a, b, c in Figure 4.3 measured along the arcs) and the three included angles at points A, B and C. The relationships between these arcs and angles are given by:

$$\begin{aligned} \cos a &= \cos b \cos c + \sin b \sin c \cos A \\ \sin a \sin B &= \sin b \sin A \\ \sin a \cos B &= \cos b \sin c - \sin b \cos c \cos A \end{aligned} \qquad (4.5)$$

4.3.2 Coordinate systems in astronomy

The terrestrial coordinate system

The Earth is a good approximation to a sphere. The axis of rotation of the Earth, termed the **polar axis**, intersects the surface of the Earth at the **north and south geographic poles**. The great circle whose plane is at right angles to the polar axis

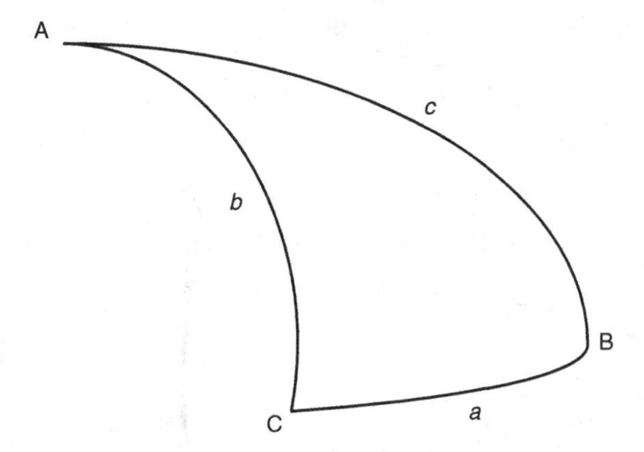

Figure 4.3. A spherical triangle.

is called the **terrestrial equator**. Other great circles that are at right angles to the terrestrial equator and pass through the north and south poles are called **terrestrial meridians**.

Geographical longitude is the angle between a terrestrial meridian circle and an arbitrarily chosen reference meridian circle called the **prime meridian**. This great circle passes through a telescope known as the **Airy transit circle** located at the Royal Observatory, Greenwich, in the United Kingdom. All longitudes are commonly measured from this prime meridian, either east or west from 0° to 180°.

Geographical latitude is the angular distance normally measured north or south of the terrestrial equator from 0° to 90° along the meridian circle through the point on the Earth's surface of interest (see Figure 4.4).

Using the radian equivalents given in Section 4.2.1 above, it is a trivial exercise to convert latitudes (ϕ) and longitudes (λ) from degrees, minutes and seconds to radians. Retaining the prime meridian in Greenwich but measuring longitude, increasing eastwards, from 0 to 2π rad, and measuring latitude from 0 at the south pole to π rad at the north pole, we may rewrite the longitude and latitude of Mauna Kea Observatory in the Hawaiian Islands (remembering that its given sexagesimal longitude is in the western hemisphere, so $\lambda = 360 - \lambda$) from 155°28′18″W; 19°49′36″N to:

$$\lambda = 3586.157\,103\,\text{mrad} \qquad \phi = 1916.836\,940\,\text{mrad}$$

Note that three decimal places corresponds to locating the observatory on the surface of the Earth to approximately ±7 m if the Earth were a perfect sphere.[12]

[12] The length of an arc of 1 radian is equal to the radius of the Earth, 6738.136 km at the Equator according to Cox (2000), so an arc length of 1 mrad = 6.738 136 km and 0.001 mrad = 6.738 m.

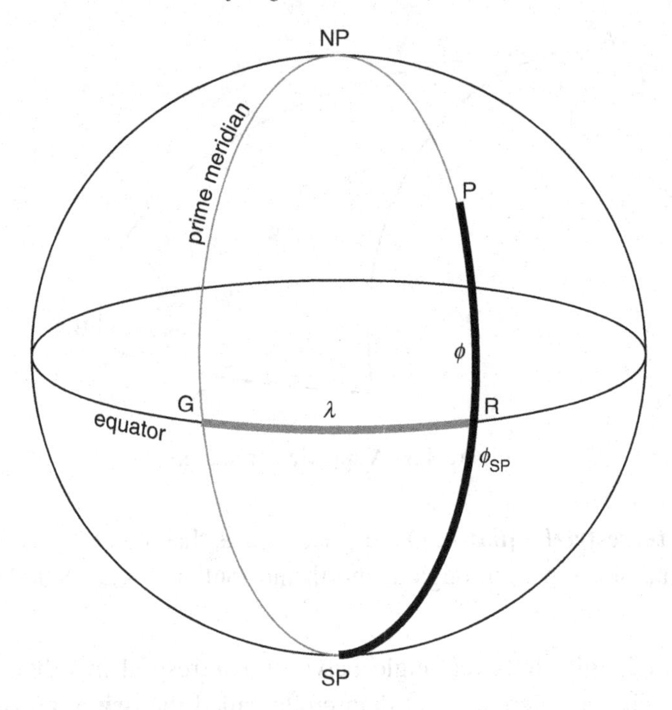

Figure 4.4. The geographical coordinates of the point P are longitude λ (the thick grey arc GR) and latitude, either ϕ (the thick black arc PR), when measured north or south from the equator, or ϕ_{SP} (the thick black arc $\overline{\text{SPRP}}$) when measured from the south pole northwards. The labelled points are: NP (north geographic pole), SP (south geographic pole), G (the intersection of the prime meridian with the equator), and R (the point of intersection of the great circle through the poles and the point P with the equator).

The horizontal or altitude–azimuth coordinate system

The great circles used in the horizontal coordinate system are the **horizon**, the projection of the terrestrial horizon of the observer to the celestial sphere and the **vertical circle**, which is at right angles to the horizon and passes through the celestial body of interest, and the **zenith**, the point in the sky directly above the observer. The vertical circle that passes through the **north** and **south** points of the horizon and the zenith is known as the **prime meridian**. The vertical circle at right angles to the prime meridian defines the **east** and **west** points where it intersects the horizon (see Figure 4.5).

The location of the celestial body is given by its **azimuth**, the angle measured along the horizon from north through east, south and west from 0 to 2000π mrad to the point of intersection of the vertical circle passing through the body, and its **altitude** (or sometimes the complement of altitude, the **zenith distance**), the angle

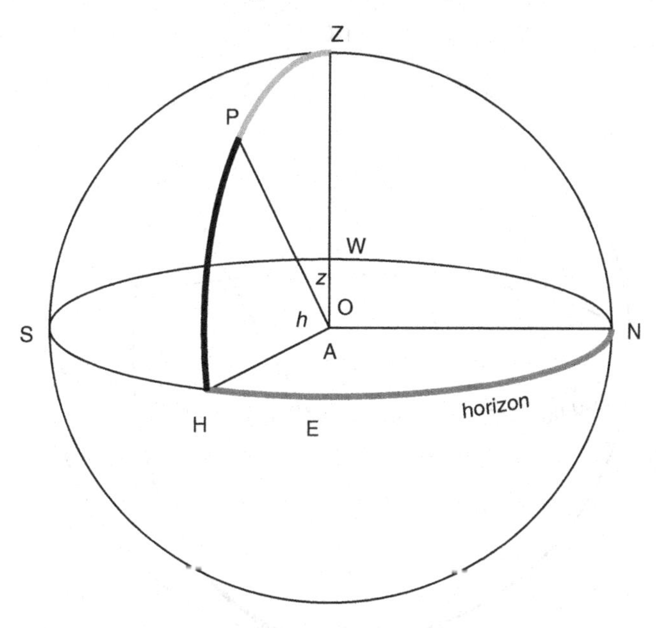

Figure 4.5. Altazimuth coordinates: azimuth angle, A is the angle NOH measured (dark grey arc) from north (N) through east (E) to the intersection point on the horizon (H) of the great circle through the zenith point (Z) and the star (P). The altitude, h, is the angle POH (thick black arc). The zenith distance, z, is the angle ZOP (light grey arc).

between the horizon and the body measured along the vertical circle through the zenith point. Hence altitude and zenith distance would lie between 0 and 500π mrad. The time and date of the observation and the location at which the observation was made must also be given, as both altitude and azimuth vary with time, date and geographical position.

The equatorial coordinate system

The equatorial system of coordinates in astronomy uses the projection of the Earth's equator on to the celestial sphere as the fundamental circle, which is known as the **celestial equator**. The projection of the Earth's north pole on to the celestial sphere is known as the **north celestial pole** (NCP in Figure 4.6) and the projection of the Earth's south pole as the **south celestial pole** (SCP in Figure 4.6). The zero point of **right ascension** is defined to be the intersection of the celestial equator with the **ecliptic** (see below) at the position where the Sun moves from south of the celestial equator to north. This point, known as the **Vernal Equinox** or the **First Point of Aries**, is point ♈ in Figure 4.6. Right ascension is measured eastwards from this point along the celestial equator.

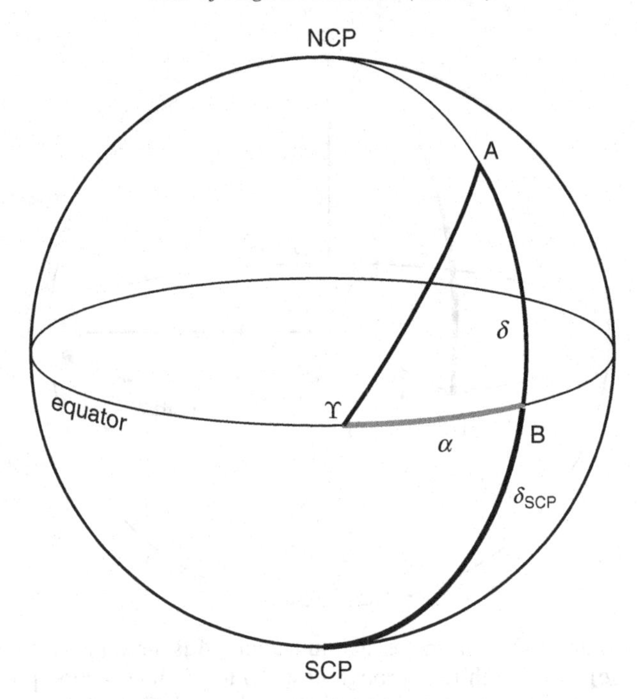

Figure 4.6. Equatorial coordinates: right ascension, α, measured from the first point of Aries (ϒ) to point B in an easterly direction (thick grey arc ϒB), declination, δ, measured from either the celestial equator at point B (thick black arc BA) or δ_{SCP}, measured from the south celestial pole northwards through B to A along the great circle through the celestial poles and point A (thick black arc $\overline{\mathrm{SCP}}$BA).

The ecliptic coordinate system

The **ecliptic** may be defined as the great circle traced out by the apparent path of the Sun across the sky during the course of one sidereal year. The ecliptic intersects the celestial equator at two points, termed the **First Point of Aries** and the **First Point of Libra** (collectively the **equinoctial points**). The angle between the celestial equator and the ecliptic is called the **obliquity of the ecliptic** and was equal to approximately 409.092 6 mrad at the standard epoch J2000.0.

Coordinates measured in the ecliptic system are **celestial longitude** (λ) and **celestial latitude** (β) (see Figure 4.7). Celestial longitude is measured eastwards from the first point of Aries along the ecliptic. Celestial latitude is measured northwards from the ecliptic (0 mrad) to the north ecliptic pole ($+500\pi$ mrad) and southwards from the ecliptic to the south ecliptic pole (-500π mrad) or alternatively, as β_{SEP}, it may be measured from 0 mrad at the south ecliptic pole (SEP) through $+500\pi$ mrad at the ecliptic and on to $+1000\pi$ mrad at the north ecliptic pole (NEP).

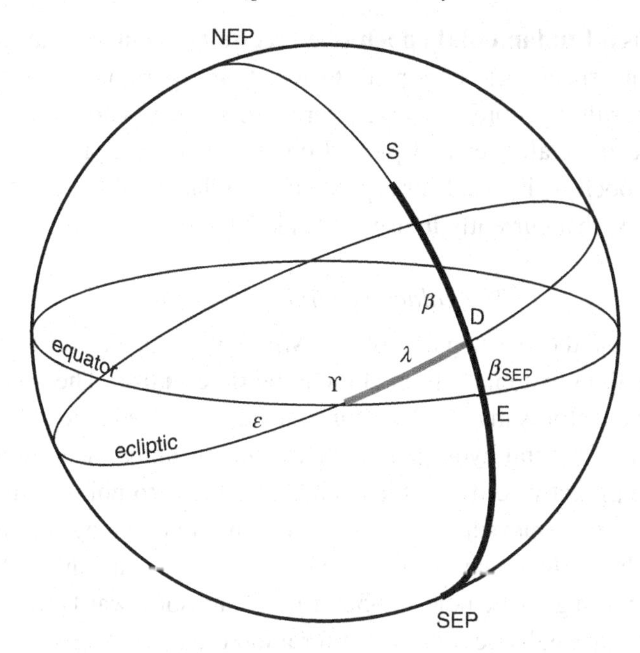

Figure 4.7. Ecliptic coordinates: celestial longitude, λ, measured from the first point of aries (Υ) eastwards along the ecliptic (thick grey arc ΥD) and celestial latitude, β, measured either north or south of the ecliptic along a great circle including the north and south ecliptic poles and the celestial object of interest, S, (thick black arc DS) or northwards from the south ecliptic pole (thick black arc $\overline{\text{SEP}}$EDS). The angle DΥE, ϵ is the obliquity of the ecliptic – the angle between the plane of the Earth's equator and its orbit about the Sun.

Disadvantages of the equatorial and ecliptic coordinate systems

The major problem with the equatorial and ecliptic coordinate systems is that their fundamental reference framework is not fixed but changes with time. Gravitational interactions between the bulge of the Earth's equator and the Sun and the Moon cause the axis of the Earth to describe a circle of radius equal to the obliquity of the ecliptic on the celestial sphere in a period of approximately 26 000 years. This effect is known as **lunisolar precession**. Perturbations from the major planets on the orbit of the Earth cause a smaller effect, known as **planetary precession**. The combination of the two is known as **general precession**. The consequence of precession is slow westward movement of the Vernal Equinox along the ecliptic at an annual rate of approximately 0.243 8 mrad (or 243.82 μrad). Superimposed on the general precessional motion is an extra movement (**nutation**) caused by the plane of the Moon's orbit rotating with respect to the ecliptic in a period of 18.6 years. When observations made over a period of years are to be compared, it is necessary for them all to be reduced to a common date or epoch. To facilitate intercomparison of observations a **fundamental epoch** (the precise moment in time for which the observations are specified) is adopted, to which all appropriate observations are reduced. The resulting coordinates are referred to as **mean coordinates**. Star positions are listed in catalogues and plotted on star charts and atlases reduced to a fundamental epoch. It is usual for a particular fundamental epoch to be used for around 50 years; that currently in use (2011) is 2000.0.

The galactic coordinate system

The projection of the mean plane of the Milky Way galaxy on to the celestial sphere defines the great circle termed the **galactic equator**. The precise location of the galactic equator is mainly determined using radio frequency observations of the distribution of neutral hydrogen gas. The same technique is used to assign the position of the **galactic centre**, which is taken as the zero point of the coordinate **galactic longitude** l, (measured eastwards from the centre along the galactic equator). **Galactic latitude** b, may be measured northwards from 0 mrad at the galactic equator to the north galactic pole at $+500\pi$ mrad and southwards from the galactic equator to the south galactic pole at -500π mrad or, alternatively, as b_{SGP}, it may be measured from 0 mrad at the south galactic pole through $+500\pi$ mrad on the galactic equator to $+1000\pi$ mrad at the north galactic pole (see Figure 4.8).

The equatorial coordinates of the galactic centre (α_{GC}, δ_{GC}) and the north galactic pole ($\alpha_{NGP}, \delta_{NGP}$) at epoch J2000.0 are given by Murray (1989) as:

$$\alpha_{GC} = 4649.644\,328\,\text{mrad} \qquad \delta_{GC} = 1065.764\,84\,\text{mrad}$$

$$\alpha_{NGP} = 3366.032\,942\,\text{mrad} \qquad \delta_{NGP} = 2044.273\,63\,\text{mrad}$$

These equatorial coordinates were calculated from the original epoch 1950.0 values agreed to by the International Astronomical Union when setting up the new system of galactic coordinates in 1958 (Blauuw *et al.*, 1960).

The values of the right ascension (α_0) and galactic longitude (l_0) of the ascending node of the galactic plane on the J2000 equator and the inclination of the galactic equator to the celestial equator (γ), derived from Cox (2000), are:

$$\alpha_0 = 4936.829\,\text{mrad}$$

$$l_0 = 574.770\,\text{mrad}$$

$$\gamma = 1097.319\,\text{mrad}$$

Although the coordinate frame in the galactic coordinate system does change with time, the effect is extremely small in comparison with either the equatorial or ecliptic coordinate systems. The galactic coordinate system is of particular value in galactic structure studies.

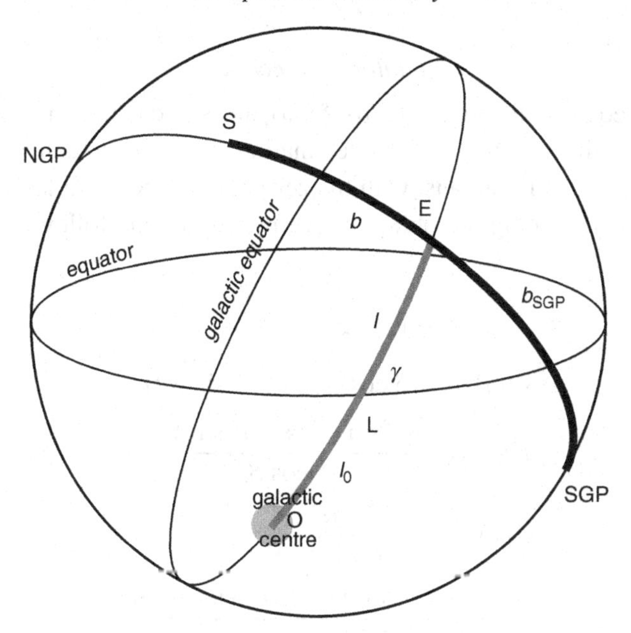

Figure 4.8. Galactic coordinates: galactic longitude, l, measured along the galactic equator from the galactic centre (dark grey arc OLE), galactic latitude, b, measured either north or south of the galactic equator (thick black arc ES) or from the south galactic pole northwards, b_{SGP}, (thick black arc $\overline{\mathrm{SGPES}}$). The longitude of the ascending node of the galactic plane, l_0, is the dark grey arc OL and the inclination of the galactic equator to the celestial equator is the angle γ.

4.3.3 Relationships between astronomical coordinate systems

Altazimuth to equatorial and vice versa

$$\begin{aligned}
\sin z \sin A &= -\cos\delta \sin H \\
\sin z \sin A &= \sin\delta \cos\phi - \cos\delta \cos H \sin\phi \\
\cos z &= \sin\delta \sin\phi + \cos\delta \cos H \cos\phi \\
\cos\delta \sin H &= \cos z \cos\phi - \sin z \cos A \sin\phi \\
\sin\delta &= \sin\phi \cos z + \cos\phi \sin z \cos A
\end{aligned} \tag{4.6}$$

where ϕ is the latitude of the observer, z is the object's zenith distance (the complement of its altitude) and H is its local hour angle related to the object's right ascension and the local sidereal time (LST) of the observation by:

$$H = \mathrm{LST} - \alpha. \tag{4.7}$$

where LST may be defined as the circle of right ascension transiting the prime meridian at the time of observation.

Equatorial to ecliptic

To convert the equatorial coordinates (α, δ_{SCP}), measured in radians, using the south celestial pole as the zero point of the declination measures, to ecliptic coordinates (λ, β), also measured in radians, with the zero point of celestial latitude being the ecliptic plane and ϵ being the obliquity of the ecliptic, the following expressions may be used:

$$\begin{aligned}
\delta &= \delta_{SCP} - \frac{\pi}{2} \\
\sin\beta &= \sin\delta\cos\epsilon - \cos\delta\sin\alpha\sin\epsilon \\
\sin\lambda &= \frac{\cos\delta\sin\alpha\cos\epsilon + \sin\delta\sin\epsilon}{\cos\beta} \\
\cos\lambda &= \frac{\cos\delta\cos\alpha}{\cos\beta} \\
\tan\lambda &= \frac{\cos\delta\sin\alpha\cos\epsilon + \sin\delta\sin\epsilon}{\cos\delta\cos\alpha} \\
\beta_{SEP} &= \beta + \frac{\pi}{2}
\end{aligned} \tag{4.8}$$

To assign the correct value for λ in four quadrants, the following conditions for the sign of each of the values of sine, cosine and tangent determine which expression is to be used to calculate the value of λ. ($\wedge$ is the Boolean symbol for 'and'.)

$$\lambda = \left\| \begin{array}{rll} \sin^{-1}\lambda & if & \sin\lambda \geq 0 \wedge \cos\lambda \geq 0 \wedge \tan\lambda \geq 0 \\ \pi - \sin^{-1}\lambda & if & \sin\lambda \geq 0 \wedge \cos\lambda < 0 \wedge \tan\lambda < 0 \\ \pi - \sin^{-1}\lambda & if & \sin\lambda < 0 \wedge \cos\lambda < 0 \wedge \tan\lambda \geq 0 \\ 2\pi + \sin^{-1}\lambda & if & \sin\lambda < 0 \wedge \cos\lambda \geq 0 \wedge \tan\lambda < 0 \end{array} \right. \tag{4.9}$$

Using as an example, the equatorial coordinates of the star Capella given above in radians, $(\alpha, \delta_{SCP}) = (1.381\,817, 2.373\,61)$, may be transformed to ecliptic coordinates using equations (4.8) and conditions (4.9) with $\epsilon = 0.409\,092\,6$ (value at J2000.0) to yield: $(\lambda, \beta) = (1.428\,690, 0.399\,06)$, where the celestial latitude is given with reference to the ecliptic; relative to the south ecliptic pole, the value would be $\beta_{SEP} = 1.979\,889$, obtained simply by adding $\pi/2$ to β.

There is perhaps a stronger case with ecliptic latitude than with equatorial declination to present the value of the coordinate with reference to the ecliptic plane, as this is of physical importance when considering Solar System bodies.

Transformation of equatorial to galactic coordinates

To convert the equatorial coordinates (α, δ_{SCP}), measured in radians, using the south celestial pole as the zero point of the declination measures, to galactic coordinates

(l, b), also measured in radians, with the zero point of galactic latitude being the galactic equator and γ being the angle of inclination between the galactic plane and the equatorial plane, the following expressions may be used:

$$\delta = \delta_{\mathrm{SCP}} - \frac{\pi}{2}$$

$$\sin b = \sin\delta\cos\gamma - \cos\delta\sin(\alpha - \alpha_0)\sin\gamma$$

$$\sin(l - l_0) = \frac{\cos\delta\sin(\alpha-\alpha_0)\cos\gamma + \sin\delta\sin\gamma}{\cos b} \tag{4.10}$$

$$\cos(l - l_0) = \frac{\cos\delta\cos(\alpha-\alpha_0)}{\cos b}$$

$$\tan(l - l_0) = \frac{\cos\delta\sin(\alpha-\alpha_0)\cos\gamma + \sin\delta\sin\gamma}{\cos\delta\cos(\alpha-\alpha_0)}$$

$$b_{\mathrm{SGP}} = b + \frac{\pi}{2}$$

where α_0 is the right ascension of the galactic centre at J2000.0 and l_0 is the galactic longitude of the point of intersection of the celestial equator and the galactic equator, known as the **longitude of the ascending node** (i.e., where $b = 0$ mrad or $b_{\mathrm{SGP}} = 1570.796$ $(= \pi/2)$ mrad).

To assign the correct value for l in four quadrants, the following conditions for the sign of each of the values of sine, cosine and tangent determine which expression is to be used to calculate the value of l. ($\wedge$ is the Boolean symbol for 'and'.)

$$l - l_0 = \left\| \begin{array}{l} \sin^{-1}(l-l_0)\ \ if\ \ \sin(l-l_0) \geq 0 \wedge \cos(l-l_0) \geq 0 \wedge \tan(l-l_0) \geq 0 \\ \pi - \sin^{-1}(l-l_0)\ \ if\ \ \sin(l-l_0) \geq 0 \wedge \cos(l-l_0) < 0 \wedge \tan(l-l_0) < 0 \\ \pi - \sin^{-1}(l-l_0)\ \ if\ \ \sin(l-l_0) < 0 \wedge \cos(l-l_0) < 0 \wedge \tan(l-l_0) \geq 0 \\ 2\pi + \sin^{-1}(l-l_0)\ \ if\ \ \sin(l-l_0) < 0 \wedge \cos(l-l_0) \geq 0 \wedge \tan(l-l_0) < 0 \end{array} \right. \tag{4.11}$$

thence

$$l = (l - l_0) + l_0 \tag{4.12}$$

Equation (4.12) can produce galactic longitudes greater than 2π, which conditional statement (4.13) removes.

$$l = \left\| \begin{array}{rl} l & \text{if}\ \ l < 2\pi \\ l - 2\pi & \text{if}\ \ l \geq 2\pi \end{array} \right. \tag{4.13}$$

Using once again as an example the equatorial coordinates of the star Capella given above as $(\alpha, \delta) = (1.381\,817\,\mathrm{rad}, 2.373\,61\,\mathrm{rad})$, which may be transformed to galactic coordinates using Equations (4.10) and conditions (4.11) with $\gamma = 1.097\,319$ rad (value at J2000.0) to yield: $(l, b) = (2.837\,704\,\mathrm{rad}, 0.079\,699\,\mathrm{rad})$, where the galactic latitude is given with reference to the galactic equator. Relative to the south galactic pole the value would be $b_{SGP} = 1.650496$ rad, obtained simply by adding $\pi/2$ to b.

The use of galactic coordinates is of greatest value when studying the distribution of galactic objects with reference to the galactic plane and centre.

4.4 Angular distances and diameters

The distance AB, between the two celestial bodies A and B shown in Figure 4.9 located at (α_A, δ_A) and (α_B, δ_B) is given in radians by:

$$\overrightarrow{AB} = \cos^{-1}(\sin\delta_A \sin\delta_B + \cos\delta_A \cos\delta_B \cos(\alpha_B - \alpha_A)) \tag{4.14}$$

If the positions of A and B are given in ecliptic or galactic coordinates, then Equation (4.12) may simply be rewritten with (λ_A, β_A) and (λ_B, β_B) or (l_A, b_A) and (l_B, b_B) substituted for (α_A, δ_A) and (α_B, δ_B).

4.4.1 Distances between pairs of astronomical objects

Select the positions of the pairs of stars α Ori and β Ori, and β Cas and β Car from Table 4.1, with declination measured from the celestial equator (i.e., $\delta = \delta_{SCP} - 500\pi$). Transform each number pair to radian measure by dividing by 1000, then substitute into Equation (4.14) above. The distance between α Ori and β Ori is 324.737 mrad. For the pair β Cas and β Car, the measured separation is 2773.878 mrad.

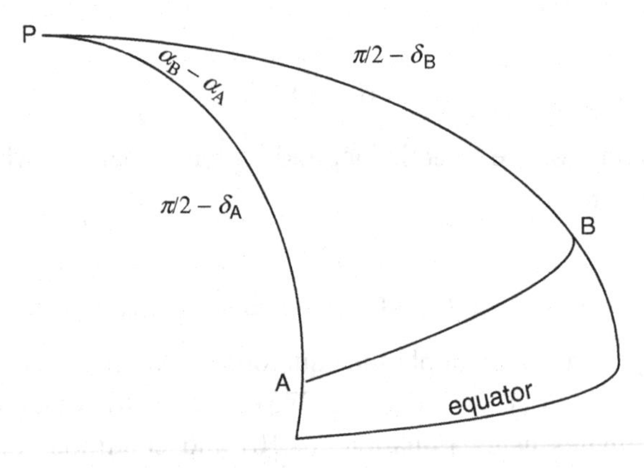

Figure 4.9. Measurement of angular distances.

Table 4.2. *Further examples of radian measures: apparent sizes and separations*

Celestial object	Sexagesimal system	Radian system	Notes, Reference
Comet Tebbutt (1861 II)	$>100°$	>1.75 rad	length of tail Chambers (1889)
LMC	$10° \, 45'.7$	187.83 mrad	galaxy Cox (2000)
M31	$3° \, 10'.5$	55.41 mrad	galaxy Cox (2000)
Pleiades	$2°$	34.91 mrad	galactic cluster Cox (2000)
Sun	$31' \, 59''.26$	9.30 mrad	star Cox (2000)
Moon	$31' \, 05''.2$	9.04 mrad	satellite Cox (2000)
Helix nebula	$12'$	3.49 mrad	planetary nebula Cox (2000)
ω Cen	$8'.36$	2.43 mrad	globular cluster Cox (2000)
Jupiter	$48''.9$	237.07 μrad	planet White (2008)
Saturn's rings	$44''.9$	217.68 μrad	planetary rings White (2008)
α Cen	$7''.53$	36.51 μrad	double star separation White (2008)
Mira	$0''.060$	290.89 nrad	stellar diameter Karovska & Sasselov (2001)
Aldebaran	$0''.020$	96.96 nrad	stellar diameter Harris *et al.* (1963)
ϵ Aur	$0''.00227$	11.01 nrad	stellar diameter Stencel *et al.* (2008)

The angular diameters or separations of a selection of celestial bodies (comet, galaxies, globular and galactic clusters, planetary nebulae, stellar diameters (including the Sun), double star separations, planets and Saturn's rings) are given in Table 4.2 in sexagesimal and radian (or subdivisions thereof) measures.

4.4.2 *Field or plate scale determination*

The plate scale of a photographic plate or a CCD is a measure of the angular distance on the celestial sphere imaged on the detector per unit linear distance measured. A common way of expressing plate scale is in arcseconds per millimetre (arcsec mm^{-1}). The equivalent SI expression is μrad mm^{-1}.

Example: determine the plate scale of the AAO Schmidt telescope (formally known as the UK Schmidt telescope)

The measured distance r, on an AAO Schmidt telescope photographic plate, between ϵ Ori and ζ Ori is:

$$r = 72.654 \, \text{mm}$$

The angular distance a between the two stars is calculated using Equation (4.14), substituting for the spherical coordinates, in radians:

$$a = \cos^{-1}[\sin\delta\sin\delta_0 + \cos\delta\cos\delta_0\cos(\alpha - \alpha_0)] = 0.023\,661\,\text{rad}$$

Hence the plate scale ($= a/r$) is:

$$\frac{a}{r} = 3.257\times 10^{-4}\,\text{rad}\,\text{mm}^{-1}$$

$$\text{or}\quad 0.326\,\text{mrad}\,.\,\text{mm}^{-1}\quad\text{or}\quad 326\,\mu\text{rad}\,.\,\text{mm}^{-1}$$

4.5 Steradian

The **steradian** is the derived SI unit of solid angle that has its apex at the centre of a sphere and subtends an area at the surface of the sphere equal to the square of the radius of the sphere.

It is a dimensionless quantity with the symbol **sr**, which may be expressed in terms of SI base units as $([L]^2\,.\,[L]^{-2})$.

The number of steradians on the entire sky is given by the following expression (4.15), where α is right ascension measured from 0 rad at the first point of Aries increasing eastwards and δ_{SCP} is declination measured northwards from 0 rad from the south celestial pole:

$$A = \int_0^{2\pi}\int_0^{\pi} \sin\delta_{SCP}\,d\,\delta_{SCP}\,d\,\alpha = 4\pi \tag{4.15}$$

Hence the number of steradians on a sphere is given by:

$$4\pi = 12.566\,371$$

4.5.1 Conversions between sexagesimal and steradian measures

The units of solid angle currently in common use in astronomy are the square degree, square arcminute and square arcsecond. To convert to SI units and vice versa, the following factors or their inverses may be used.

The number of square degrees in a sphere:

$$4\pi\left(\frac{180}{\pi}\right)^2 = 41\,252.961\,249$$

the number of square degrees in a steradian:

$$\left(\frac{180}{\pi}\right)^2 = 3282.806\,35$$

one square degree in steradians:

$$\left(\frac{\pi}{180}\right)^2 = 0.000\,304\,617\,\text{sr} = 304.617\,\mu\text{sr}$$

one square arcminute in steradians:

$$\left(\frac{\pi}{180 \times 60}\right)^2 = 8.461\,595 \times 10^{-8}\,\text{sr} = 84.615\,95\,\text{nsr}$$

one square arcsecond in steradians:

$$\left(\frac{\pi}{180 \times 3600}\right)^2 = 2.350\,443 \times 10^{-11}\,\text{sr} = 23.504\,43\,\text{psr}$$

4.5.2 Area of the constellation Crux

The IAU boundaries of the constellation Crux are defined by two lines of constant declination and two of constant right ascension.

The approximate equatorial coordinates in radians of the northernmost (δ_{max}) = 0.598, southernmost (δ_{min}) = 0.441, easternmost (α_{max}) = 3.393 and westernmost (α_{min}) = 3.128 points were determined from the digital sky atlas *The Sky*™.

The area of the constellation Crux is then derived by evaluating the integral:

$$A_{\text{Crux}} = \int_{3.128}^{3.393} \int_{0.441}^{0.598} \sin\delta\, d\,\delta\, d\,\alpha = 0.0206\,\text{sr} \tag{4.16}$$

In sexagesimal units, this computed area is 67.7 square degrees, in tolerable agreement with the value of 68 square degrees given by Moore (2001).

For constellations with boundaries that follow more than two circles of right ascension and/or declination, a more general formulation for the total area A_{tot} would be:

$$A_{\text{tot}} = \sum_{i=0}^{n} \int_{\alpha_{\text{min}_i}}^{\alpha_{\text{max}_i}} \int_{\delta_{\text{min}_i}}^{\delta_{\text{max}_i}} \sin\delta\, d\,\delta\, d\,\alpha \tag{4.17}$$

4.5.3 Further examples of angular area measurement in astronomy

Example: determine the angular area of the Sun and Moon

From Table 4.2 above, the mean apparent angular radius of the Sun is 4.65 mrad. Assuming the Sun to be spherical, its mean apparent angular area $A_\odot$ is

$$\begin{aligned} A_\odot &= \pi(4.65)^2 \\ &= 67.93\,\text{mrad}^2 \text{ or } 67.93\,\mu\text{sr} \end{aligned} \tag{4.18}$$

For the Moon, the mean apparent angular radius derived from Table 4.2 is 4.57 mrad and hence its area is

$$A_{\text{Moon}} = 65.61\,\text{mrad}^2 \;\text{ or }\; 65.61\,\mu\text{sr} \tag{4.19}$$

Example: determine the visible angular area of the planet Venus on 2009 February 28

On 2009 February 28, the angular radius of the planet Venus was 107.6 μrad and the percentage, p, of the disc illuminated by the Sun was 20%. Hence, the angular area of the illuminated portion of the disc was:

$$\begin{aligned} A_{\text{Venus}} &= p\pi(107.6)^2 \\ &= 7278\,\mu\text{rad}^2 \;\text{ or }\; 7\,278\,\text{psr} \end{aligned} \tag{4.20}$$

Example: determine the angular area of the disc of the F supergiant star ϵ Aurigae

The recently measured radius (Stencel *et al.*, 2008) of the F-type supergiant primary of the eclipsing binary star ϵ Aurigae determined using the Palomar Testbed Interferometer was 5.5 nrad. Assuming a uniform stellar disc, the corresponding apparent angular area of the ϵ Aur primary is:

$$\begin{aligned} A_{\epsilon\text{Aur}} &= \pi\,(5.5)^2 \\ &= 95\,\text{nrad}^2 = 95\,\text{asr}\;\;\text{(attosteradians)} \end{aligned} \tag{4.21}$$

Example: determine the angular area of the Local Group Galaxy M31

The measured angular diameter of M31 is given (Cox, 2000) as 8.7 arcmin or 522 arcsec within the boundary set by the B = 25 mag . arcsec^{-2} isophote, and the axial ratio $\frac{b}{a}$ as 0.32. The area (πab) of a fitted ellipse with semi-major axis a and semi-minor axis b is:

$$\begin{aligned} A &= \pi\,ab \\ A &= 6.848 \times 10^4\,\text{arcsec}^2 \\ &= 1.610 \times 10^{-6}\,\text{sr} \\ &= 1.610\,\mu\text{sr} \\ &= 1.610\,\text{mrad}^2 \end{aligned} \tag{4.22}$$

To convert the isophote unit from magnitudes per arcsecond squared to janskys per steradian, the following method may be used: the B = 25 mag . arcsec^{-2} is 10^{-10} times fainter than the B = 0 mag . arcsec^{-2} isophote. B = 0 is equivalent to a monochromatic flux density of 4000 Jy (Bessell, 1992), so B = 25 is equivalent to a monochromatic flux density of 400 nJy (4×10^{-7} Jy). Using the conversions between arcsec^2 and sr given above, the monochromatic flux density per steradian is:

$$\begin{aligned}\frac{4\times 10^{-7}}{2.350443\times 10^{-11}} &= 17018\,\mathrm{Jy\,.\,sr^{-1}}\\ &= 1.7018\times 10^{-8}\,\mathrm{Jy\,.\,psr^{-1}}\\ &= 1.7018\,\mathrm{Jy\,.\,\mu rad^{-2}}\end{aligned} \tag{4.23}$$

So the apparent angular area of M31 to the limiting isophote of 17018 Jy . sr^{-1} is 1.610×10^{-6} sr.

Number densities of galaxies

Counting the numbers of objects in cells of similar shape and size has been used to determine the large-scale distribution of astronomical objects for more than half a century. The general luminosity function of stars (number of stars per unit brightness interval per unit area) was first studied in detail by Bok (1937). Some 20 years later, Shane & Wirtanen (1954) counted the number of galaxies on photographic plates in unit areas to identify clusters of galaxies.

As an example of counting galaxies in cells, an area of 256 μsr (16 mrad × 16 mrad), centred on the cluster of galaxies A3266 was measured using the COSMOS automated photographic-plate-measuring machine at the Royal Observatory Edinburgh (Longair, 1989). The equatorial coordinates for the centre of A3266 at J2000.0 are approximately:

$$\alpha = 1183\,\mathrm{mrad} \qquad \delta_{\mathrm{SCP}} = 498\,\mathrm{mrad}$$

For each image detected by COSMOS, 13 parameters relating to the image centroid, its size, brightness, orientation and shape were measured, which enabled separate lists of galaxy images, star images and non-standard images to be compiled. In total, 4328 images were measured, of which 1256 were classified as galaxies and three as bright galaxies. The field was divided into a grid of 256 (16 × 16) equal-sized (1 mrad × 1 mrad = 1 μsr) areas, and the number of galaxies in each counted (see Table 4.3). These numbers were processed by a contour-plotting routine, the output of which is shown in Figure 4.10.

The mean of the cell count numbers and the standard deviation about that mean are 4.3 ± 3.9 galaxies per cell, the median cell count is 3.5 galaxies per cell and the mode is 2 galaxies per cell.

Table 4.3. *Cell counts (per mrad^2) for galaxies in the neighbouring field and cluster of galaxies A3266*

0	4	1	1	0	3	4	6	2	0	3	3	6	3	2	6
5	4	6	4	0	3	2	2	4	2	2	1	8	6	4	2
3	1	4	2	0	1	1	5	4	2	0	8	4	11	17	3
2	2	2	2	3	2	1	4	2	1	3	4	4	4	6	18
1	1	1	7	3	3	5	8	4	4	2	12	10	3	7	1
2	3	3	1	6	2	7	5	8	8	11	5	9	8	2	0
4	3	7	2	4	7	7	16	7	5	7	2	6	8	3	2
2	0	3	5	1	9	8	27	20	8	6	7	13	9	0	1
3	3	3	5	3	1	5	13	27	12	5	4	6	3	6	4
4	3	0	2	4	7	8	8	17	7	17	2	2	6	4	4
1	4	1	2	2	2	6	5	9	5	5	7	9	5	2	3
5	2	3	3	0	3	3	3	7	9	6	4	3	4	2	4
2	1	1	4	1	5	2	3	2	8	1	6	5	6	1	5
5	3	2	1	4	5	0	5	6	4	3	2	4	1	2	1
4	3	6	4	1	4	2	1	2	4	2	3	1	1	3	1
4	4	2	0	2	5	0	5	5	4	0	2	4	3	2	5

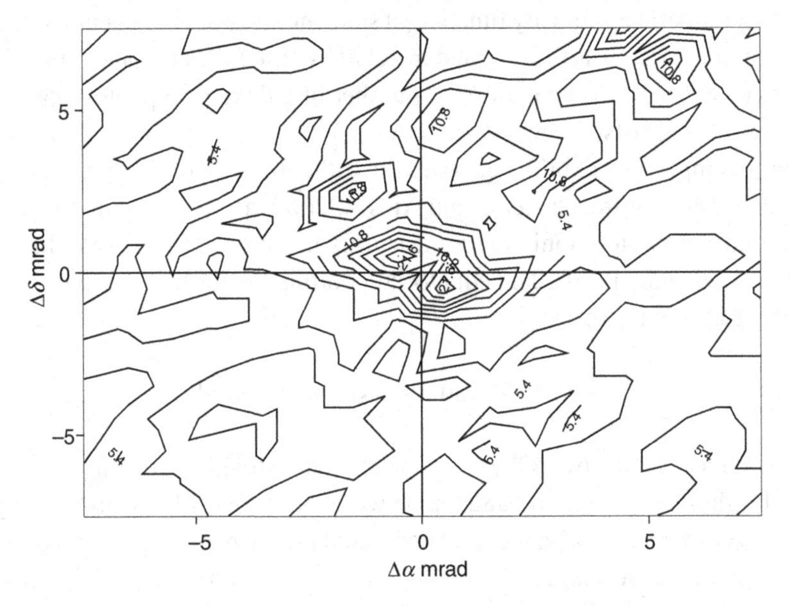

Figure 4.10. Number of galaxies per microsteradian, from data obtained using the COSMOS measuring machine, of the cluster of galaxies A3266 and its surrounding area from which all the star images have been removed.

Figure 4.10 clearly shows that A3266 is in fact two adjacent clusters of galaxies and not just one, which Henriksen & Tittley (2002), using the X-ray space-based CHANDRA Observatory, have recently shown to be in the process of merging.

4.6 Summary and recommendations

4.6.1 Summary

The SI unit of angular measure is the radian. The units commonly used by astronomers to define the positions of celestial objects are hours, minutes and seconds for the right ascension coordinate and degrees, minutes and seconds for the declination coordinate. Transformations from these units to radians and recommended submultiples of radians are given. The common types of astronomical coordinate systems are described and the relationships linking them given. The coordinates of different types of astronomical bodies, their angular distances apart and their angular sizes and areas are presented in tables and by way of worked examples.

4.6.2 Recommendations

The ease with which the location of points on the terrestrial and celestial spheres may be made is mentioned, with a pair of numbers replacing the current six. The step from sexadecimal to radian measure is considered too large by some astronomers so, as a first step in this process, using decimal degrees rather than degrees, minutes and seconds may make the final changeover to radians easier.

5

Unit of time (second)

5.1 SI definition of the second

The **second** is the duration of 9 192 631 770 periods of the radiation corresponding to the transition between the two hyperfine levels of the ground state of the caesium 133 atom.

This definition refers to a caesium atom in its ground state at a temperature of 0 K. The dimension of time is **[T]**, its unit is the **second** and its symbol is **s**.

5.2 Definition of time

Funk & Wagnalls New Standard Dictionary of the English Language (Funk *et al.*, 1946) defines time, *inter alia*, as:

> The general idea, relation, or fact of continuous or successive existence; or the abstract conception of duration as limitless, capable of division into measurable portions, and essentially comprising the relations of present, past and future.

A system of reckoning or measuring duration; as solar *time*; sidereal *time*; mean *time*.

5.3 Systems of time or time scales

There are two major systems of time: those based on the Earth's rotation and the orbital motions of the Earth, Moon and planets, known as **dynamical time**; and those based on atomic clocks[13] and known as **atomic time**.

A time system or scale may be specified by two numbers, the origin from which the time intervals are to be measured and the number of predefined unit scale intervals measured since the time of origin (Leschiutta, 2001).

[13] See, e.g., http://tycho.usno.navy.mil/clockdev/cesium.html

5.3.1 Dynamical time

Dynamical time may be thought of as the independent variable in the equations that describe the motions of the bodies in the Solar System. It was developed from the natural system of timekeeping based on the apparent motion of the Sun across the celestial sphere (see, e.g., Lang (2006), Cox (2000) and the USNO website[14] on which the greater part of this section is based).

5.3.2 Atomic time

By the middle of the twentieth century the relative accuracy of the unit dynamical time interval was, at best, around 1 part in 10^8, which was proving inadequate for particular applications in astronomy, spectroscopy and telecommunications. This accuracy limit was imposed by the irregularities in the period of the Earth's rotation. The continued developments of atomic clocks in the 1950s led to an improvement in the accuracy of time measurement to 1 part in 10^9, which in turn led to the discovery of the variability in the Earth's rotation period. Currently, the most accurate atomic clock is a type known as a caesium fountain atomic clock, with a measurement uncertainty of around 5 parts in 10^{16}.[15]

5.3.3 Time systems currently in use

Atomic time is defined above in terms of the duration of a specified number of cycles of radiation corresponding to the transition of two hyperfine levels of the ground state of ^{133}Cs at absolute zero.

TAI, International Atomic Time, is calculated from a statistical analysis of individual frequency standards and time scales based on atomic clocks situated throughout the world.

UT, Universal Time, based on the mean solar day, is the time system used for all civil timekeeping. UT0 is rotational time for a particular location on the Earth, uncorrected for shifts in longitude due to polar motion. UT1 is the rotational time corrected for such shifts, though it is still non-uniform, being subject to the irregularities in the Earth's rotation.

UTC, Coordinated Universal Time, is the time scale distributed by means of radio time signals, satellites, radio and TV broadcasts. UTC differs from TAI by an

[14] www.usno.navy.mil/USNO/time/master-clock/systems-of-time
[15] See http://tycho.usno.navy.mil/clockdev/cesium.html

integral number of seconds and is kept to within 0.9 s of UT1 by means of the irregular introduction of integer leap seconds. UTC has replaced GMT (Greenwich Mean Time) as an international time standard.

TT or TDT, Terrestrial Dynamic Time, is the idealized time based on the rotation period of the geoid. The time unit is 1 day consisting of 86 400 s in SI units. It is approximately related to TAI by the expression:

$$\mathrm{TDT} = \mathrm{TAI} + 32.184\,\mathrm{s} \tag{5.1}$$

TDB, Barycentric Dynamical Time, is the relativistically transformed time for referring equations of motion to the barycentre of the Solar System. It is defined to differ from TDT solely by periodic variations such that:

$$\mathrm{TDB} = \mathrm{TDT} + 0^{\mathrm{s}}.001\,658 \sin g + 0^{\mathrm{s}}.000\,014 \sin 2g \tag{5.2}$$

$$g = 357^\circ.53 + 0^\circ.985\,600\,28\,(\mathrm{JD} - 2\,451\,545.0) \tag{5.3}$$

$$= 6.240\,075\,675 + 0.017\,201\,969\,(\mathrm{JD} - 2\,451\,545.0)\,\mathrm{rad} \tag{5.4}$$

TCG, Geocentric Coordinate Time, is a coordinate time related to the centre of mass of the Earth as its spatial origin. It is related to TDT by:

$$\mathrm{TCG} = \mathrm{TDT} + 6.969\,2904 \times 10^{-10}\,(\mathrm{JD} - 2\,443\,144.5) \times 86\,400\,\mathrm{s} \tag{5.5}$$

where JD is the julian day number equal to the number of days that have elapsed since Greenwich Mean Noon on 1 January 4713 BCE, Julian proleptic calendar (the Julian Date is the julian day plus the elapsed time since the preceding noon).

TCB, Barycentric Coordinate Time, is a coordinate time related to the barycentre of the Solar System as its spatial origin. TCB differs from TDB in rate and is related to it by:

$$\mathrm{TCB} = \mathrm{TDB} + 1.550\,506 \times 10^{-8}\,(\mathrm{JD} - 2\,443\,144.5) \times 86\,400\,\mathrm{s} \tag{5.6}$$

ST, Sidereal Time, is defined to be the hour angle of the First Point of Aries, ♈, the point of intersection of the celestial equator and the ecliptic (see Chapter 4). In other words, the right ascension in transit over the prime meridian (the north–south line) at a given terrestrial longitude (λ). The relationship between the local sidereal time (LST) and the Greenwich sidereal time (GST) is:

$$\mathrm{LST} = \mathrm{GST} + \lambda \tag{5.7}$$

where λ is measured eastwards from Greenwich. The local hour angle (LHA) of a celestial object S at right ascension, α_S, is:

$$LHA_S = LST - \alpha_S \tag{5.8}$$

The date of any such measurement of the local hour angle has to be given, as the location of ♈ is not fixed due to the effects of precession and nutation.

MJD, Modified Julian Date, which begins and ends at midnight, is defined in terms of the julian day number (see above) as:

$$MJD = JD - 2\,400\,000.5 \tag{5.9}$$

For precise applications the time scale should be specified.

5.3.4 Multiples of the second

Multiples of the second in common use are as follows:

1 minute = 60 s
1 hour = 60 m = 3 600 s
1 day = 24 h = 1 440 m = 86 400 s

5.3.5 Leap second

In order to ensure that the difference between time determined using atomic clocks and that determined from the Earth's rotation does not exceed 0.9 s, civil time (UTC) is occasionally adjusted in one second increments.

An example of the application of the leap second is the sequence of dates created at the end of December 2008:[16]

2008 December 31,	23 h 59 m 59 s
2008 December 31,	23 h 59 m 60 s
2009 January 01,	00 h 00 m 00 s

The difference (UTC−TAI) applied to TAI was:

−33 s from 2006 January 01, 0 h UTC to 2009 January 01, 0 h UTC;
−34 s from 2009 January 01, 0 h UTC until further notice.

[16] See http://tycho.usno.navy.mil/bulletinc2008.html

5.3.6 Relationships between mean solar time and mean sidereal time

The mean sidereal rotation period of the Earth is equivalent to (*The Astronomical Almanac, 1995*):

86 164.090 54 s (SI units) or
23 h 56 m 04.090 53 s (mean solar time) or
0.997 269 566 33 d (mean solar time)

One mean solar rotation period of the Earth is equivalent to:

1.002 737 909 35 d (mean sidereal time) or
24 h 03 m 56.555 37 s (mean sidereal time)

These values relate to the year 1995.

5.3.7 The month

The length of the month depends on the reference point chosen to define the start and end of the period. Table 5.1 gives the lengths of the mean months for 1995.0 (*The Astronomical Almanac, 1995*).

5.3.8 The year

The Julian year is defined to be 365.25 d precisely and forms the basis for the Julian calendar. The length of the Julian year may also be written as:

$$1\ \mathrm{y} = 365.25\ \mathrm{d} = 8\,766\ \mathrm{h} = 525\,960\ \mathrm{m} = 31\,557\,600\ \mathrm{s}\ \text{(SI unit)}.$$

The lengths of some other types of year are given in Table 5.2.

5.3.9 ISO8601 standard on dates and time

The International Organization for Standardization (ISO) has prepared a standard, ISO8601, which sets out a method for numerically representing the following:

Date: Time of the day: Coordinated Universal Time (UTC): Local time with offset to UTC.
Date and time: Time intervals: Recurring time intervals.

Formats of particular use in astronomy are:

1. Calendar date: YYYY–MM–DD, where YYYY is the Gregorian calendar year, MM is the month of the year (01 = January, 02 = February, 03 = March … 12 = December) and DD is the day of the month from 01 to 28, 29, 30 or 31.
   For example, 1944–12–08 is the 8th December 1944.

Table 5.1. *Name, definition and length of types of month*

Month type Reference point	length d	d	length h	m	s	length s
Draconian Node	27.212 221	27	05	05	35.9	2 351 135.9
Tropical Equinox	27.321 582	27	07	43	04.7	2 360 584.7
Sidereal Fixed star	27.321 662	27	07	43	11.6	2 360 591.6
Anomalistic Perigee	27.554 550	27	13	18	33.1	2 380 713.1
Synodic New Moon	29.530 589	29	12	44	02.9	2 551 442.9

Table 5.2. *Name, definition and length of named years at 1995.0*

Year type, Reference point	length d	d	length h	m	s	length s
Eclipse Node	346.620 074	346	14	52	54.4	29 947 974.4
Tropical Equinox	365.242 190	365	05	48	45.2	31 556 925.2
Gregorian Gregorian calendar	365.242 5	365	05	49	12	31 556 952
Julian Julian calendar	365.25	365	06	0	0	31 557 600
Sidereal Fixed star	365.256 36	365	06	09	10	31 558 150
Gaussian Kepler's law for $a = 1$	365.256 90	365	06	09	56	31 558 196
Anomalistic Perihelion	365.259 64	365	06	13	53	31 558 433

2. Time of the day uses the 24-hour system in the form hh:mm:ss starting with 00 for the first hour after midnight and ending with 23 as the last before the following midnight. Hence, 21:56:16 is equivalent to 9 h 56 m 16 s p.m. or 3 m 44 s before 10 p.m.
3. The representation of both date and time as a single composite number uses the capital letter T to separate the date component from the time component.

For example, 2003–09–03 T 05:21:37 is 21 m 37 s after 5 on the morning of 3rd September 2003.[17]

5.4 The hertz: unit of frequency

The derived SI unit of **frequency** is the hertz. The dimension of frequency is $[\boldsymbol{T}]^{-1}$, its unit is the **hertz** and its symbol is **Hz**. The unit was previously known as the 'cycle per second', $\mathrm{c\,.\,s^{-1}}$.

5.5 Angular motion

5.5.1 Angular velocity and acceleration

The SI derived unit of angular velocity has dimension $\mathbf{[1]}\,.\,[\boldsymbol{T}]^{-1}$, its unit is the **radian per second** and its symbol is $\mathbf{rad\,.\,s^{-1}}$.

The angular velocity, ω, may be expressed as the first derivative with respect to time of the angle θ, thus:

$$\omega = \frac{\mathrm{d}\theta}{\mathrm{d}t} = \dot{\theta} \tag{5.10}$$

Angular acceleration, $\dot{\omega}$, is the second derivative with respect to time of the angle θ, thus:

$$\dot{\omega} = \frac{\mathrm{d}^2\theta}{\mathrm{d}t^2} = \ddot{\theta} \tag{5.11}$$

The SI derived unit of angular acceleration has a dimension of $\mathbf{[1]}\,.\,[\boldsymbol{T}]^{-2}$, its unit is the **radian per second per second** and its symbol is $\mathbf{rad\,.\,s^{-2}}$.

5.5.2 Rotation period and period of revolution

The difference between rotation and revolution is defined as (Funk *et al.*, 1946) 'To rotate is said of a body that has a circular motion about its own centre or axis; to revolve is said of a body that moves about a centre outside of itself.'

Some examples of angular velocities, rotation periods and periods of revolution for Solar System bodies

For planets and other bodies in the Solar System, the mean orbital motion is often quoted in units of degrees per day. In SI units, $\mathrm{rad\,.\,s^{-1}}$ is used. The conversion from degrees per day to $\mathrm{rad\,.\,s^{-1}}$ is carried out as follows:

$$1^\circ.\mathrm{d}^{-1} = \frac{\pi}{180 \times 86400}\,\mathrm{rad\,.\,s^{-1}} = 2.020057005 \times 10^{-7}\,\mathrm{rad\,.\,s^{-1}} \tag{5.12}$$

[17] For further information about ISO8601, see the ISO website at: www.iso.org/iso/support/faqs/

Table 5.3. *Rotation and revolution periods and rates for the planets and some dwarf planets in the Solar System*

Planet	Sidereal period Julian years	Mean orbital angular velocity nrad .s^{-1}	Sidereal rotation rate days	Rotational angular velocity μrad .s^{-1}
Mercury	0.240 844	826.683 495	58.646 225	1.240 013
Venus	0.615 182	323.647 219	−243.019 99	−0.299 243
Earth	0.999 978	199.106 385	0.997 269	72.921 154
Mars	1.880 711	105.865 348	1.025 957	70.882 181
Jupiter	11.856 525	16.792 621	0.413 538	175.853 234
Saturn	29.423 519	6.766 769	0.444 009	163.784 990
Uranus	83.747 407	2.377 413	−0.718 333	−101.237 196
Neptune	163.723 204	1.216 090	0.671 250	108.338 253
Dwarf planet				
Ceres	4.60	43.283 071	0.378 125	192.322 783
Pluto	248.020 8	0.802 764	−6.387 246	−11.385 510
Eris	560.89	0.354 974	≥ 0.583 ?	≤ 124.666 ?

Table 5.3 gives the sidereal period of revolution in julian years, the sidereal rotation period in days, the mean orbital angular velocity in nrad . s^{-1} and the rotational angular velocity in μrad . s^{-1}.

The negative values for the mean rotational periods and angular motions of the planets Venus and Uranus and the dwarf planet Pluto are due to their retrograde motions. The sources used for Table 5.3 were: Cox (2000); *The Astronomical Almanac* (1995); Duffard *et al.* (2008) and the JPL Small-Body Database Browser.[18]

5.5.3 Proper motions

The proper motion of a celestial body may be defined as the change in direction of that object across the celestial sphere over a period of time, as seen by an observer located at the Sun.

This total proper motion is the sum of the actual motion of the celestial body (its peculiar motion) and the parallactic motion, which is the reflex motion of the Sun about the galactic centre.

In Figure 5.1 the celestial body is initially located at S_0, with equatorial coordinates (α_0, δ_0) and declination measured from the celestial equator, north positive and south negative. The body has a proper motion of magnitude μ, and direction of motion θ, measured in a direction from the north through east. After a time interval t, measured in julian years, the body moves to a new position S_1 with

[18] See http://ssd.jpl.nasa.gov/sbdb.cgi

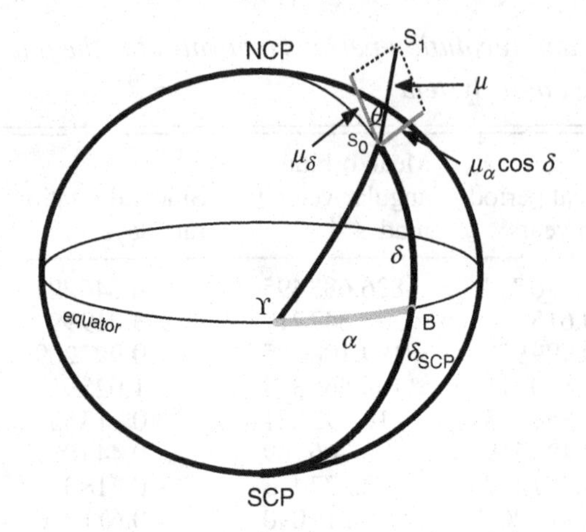

Figure 5.1. The proper motion, over a period of time t years, of the celestial body at S_0, with equatorial coordinates (α_0, δ_0), takes it to position S_1 with coordinates (α_1, δ_1).

equatorial coordinates (α_1, δ_1) relative to the Sun. The components of the annual proper motion in the equatorial coordinate system are (μ_α, μ_δ) where:

$$\mu_\alpha = \frac{\alpha_1 - \alpha_0}{t}$$

and

$$\mu_\delta = \frac{\delta_1 - \delta_0}{t} \tag{5.13}$$

and the magnitude, μ, is given by:

$$\mu = \sqrt{\mu_\alpha^2 \cos^2 \delta + \mu_\delta^2} \tag{5.14}$$

the factor $\cos\delta$ being due to the radius of the small circle of declination being a function of δ. The position angle θ, as shown in Figure 5.1, is related to the proper-motion components by:

$$\mu \cos\theta = \mu_\delta$$

$$\mu \sin\theta = \mu_\alpha \cos\delta \tag{5.15}$$

Proper motions in SI units

The SI unit of proper motion has a dimension of **[1]** . $[T]^{-1}$, its unit is the **radian per second** and its symbol is **rad . s**$^{-1}$.

Units in common use include the arcsecond per Julian year (arcsec . y^{-1}); the arcsecond per Julian century (arcsec . $(\text{century})^{-1}$); the milliarcsecond per Julian year ($\mathrm{mas\,.\,y^{-1}}$) and by using radio interferometry, the microarcsecond per Julian year ($\mathrm{\mu as\,.\,y^{-1}}$).

Converting commonly used proper-motion units into SI units

$$1\,\mathrm{arcsec\,.\,y^{-1}} = \frac{\pi}{180 \times 3600 \times 31.5576 \times 10^{6}}\,\mathrm{rad\,.\,s^{-1}}$$

$$= 1.536\,281\,85 \times 10^{-13}\,\mathrm{rad\,.\,s^{-1}} \tag{5.16}$$

$$1\,\mathrm{mas\,.\,y^{-1}} = 1.536\,281\,85 \times 10^{-16}\,\mathrm{rad\,.\,s^{-1}} \tag{5.17}$$

$$1\,\mathrm{\mu as\,.\,y^{-1}} = 1.536\,281\,85 \times 10^{-19}\,\mathrm{rad\,.\,s^{-1}} \tag{5.18}$$

$$1\,\mathrm{arcsec\,.\,y^{-1}} = \frac{\pi}{180 \times 3\,600}\,\mathrm{rad\,.\,y^{-1}}$$

$$= 4.848\,136\,811 \times 10^{-6}\,\mathrm{rad\,.\,y^{-1}} \tag{5.19}$$

$$1\,\mathrm{arcsec\,.\,(century)^{-1}} = 4.848\,136\,811 \times 10^{-6}\,\mathrm{rad\,.\,(century)^{-1}} \tag{5.20}$$

Table 5.4 gives the proper motions of the ten brightest stars in the night sky, taken from values obtained by the HIPPARCOS astrometric satellite and listed on the ESO website database.[19] The basic SI unit of angular velocity (column 3 in Table 5.4) could be rendered in alternative forms as, e.g., $\mathbf{-83.88\,frad\,.\,s^{-1}}$ (femtoradians per second) or $\mathbf{-83.88}$ **quadrillionths** $\mathbf{rad\,.\,s^{-1}}$ or $\mathbf{-8.388\,d\,14\,rad\,.\,s^{-1}}$, which is equivalent to $\mathbf{-8.388 \times 10^{-14}\,rad\,.\,s^{-1}}$.

5.5.4 Proper-motion catalogues

The introduction of high-speed measuring engines, astrometric satellites, orbiting observatories, large ground-based telescopes with adaptive optics and smaller telescopes dedicated to a specific project (not necessarily astrometric) has led to a veritable explosion of astrometric and photometric data available online. Three examples of such catalogues with sample entries in both the original data format and SI format follow.

The HIPPARCOS/TYCHO satellite provided a large body of positional, distance and proper-motion data without the restrictions placed by the Earth's atmosphere on ground-based observations. The entire database was recently revised by van Leeuwen (2007). The database is restricted to stars whose monochromatic flux density in the B waveband exceeds approximately 40 mJy (B magnitude $\sim$12.5 in the catalogue). The median precision of the proper motions is $12\,\mathrm{nrad\,.\,y^{-1}}$ ($2.5\,\mathrm{mas\,.\,y^{-1}}$). The database has been of great importance in establishing the

[19] See http://archive.eso.org/skycat/servers/ASTROM

Table 5.4. *Proper motions of the ten brightest stars in the night sky*

Star name	$\mu_\alpha \cos\delta$ μ_δ mas . y^{-1}	$\mu_\alpha \cos\delta$ μ_δ rad . s^{-1}	$\mu_\alpha \cos\delta$ μ_δ nrad . y^{-1}
Sirius	−546.01	-8.388×10^{-14}	−2647.13
	−1223.08	-1.879×10^{-13}	−5929.66
Canopus	19.99	3.071×10^{-15}	96.91
	23.67	3.636×10^{-15}	114.76
α Cen A	−3678.19	-5.651×10^{-13}	17832.37
	481.84	7.402×10^{-14}	2336.03
Arcturus	−1093.45	-1.680×10^{-13}	−5301.20
	−1999.40	-3.072×10^{-13}	−9693.36
Vega	201.02	3.088×10^{-14}	974.57
	287.46	4.416×10^{-14}	1393.65
Capella	75.52	1.160×10^{-14}	366.13
	−427.13	-6.562×10^{-14}	−2070.78
Rigel	1.87	2.873×10^{-16}	9.07
	−0.56	-8.603×10^{-17}	−2.71
Procyon	−716.57	-1.101×10^{-13}	−3474.03
	−1034.58	-1.589×10^{-13}	−5015.79
Betelgeuse	27.33	4.199×10^{-15}	132.50
	10.86	1.668×10^{-15}	52.65
Achernar	88.02	1.352×10^{-14}	426.73
	−40.08	-6.157×10^{-15}	−194.31

Table 5.5. *HD73904, a member of the galactic cluster IC2391; data from the revised HIPPARCOS/TYCHO catalogue by van Leeuwen (2007)*

			α	δ	μ_α	μ_δ
Unit	HD	HIP	h m s	° ′ ″	mas . y^{-1}	mas . y^{-1}
ID	number	number	mrad	mrad	nrad . y^{-1}	nrad . y^{-1}
IAU	73904	42374	08 38 23.94	−53 43 18.6	−23.49	21.86
SI			2261.942	−937.623	−113.88	105.98

International Celestial Reference Frame (McCarthy & Petit, 2004). Examples of converting the catalogue proper motions in mas . y^{-1} to SI units are given in Tables 5.4. and 5.5.

In Table 5.5, the unit ID ‘IAU’ gives the positions and proper motions in what are the most commonly used and accepted forms at present. It should be noted that in the van Leeuwen (2007) revision of the HIPPARCOS/TYCHO catalogue both right

Table 5.6. *A partial entry for the star id13222 from the OGLE-II proper-motion catalogue by Sumi et al. (2004)*

Unit	α	δ	μ_α	μ_δ	μ_l	μ_b
	°	°	mas y^{-1}	mas y^{-1}	mas y^{-1}	mas y^{-1}
IAU	271.010 58	−29.241 83	−3.98	−6.50	−7.62	0.31
	mrad	mrad	nrad . y^{-1}	nrad . y^{-1}	nrad . y^{-1}	nrad . y^{-1}
SI	4730.027	−510.366	−19.30	−31.51	−36.94	1.50

ascension and declination are presented in radians, though not the proper-motion components.

On a much larger scale is the USNO B1.0 all sky catalogue of magnitudes, colours, positions and proper motions for more than one billion celestial objects to a limiting monochromatic flux density of 14.3 μJy in the V waveband (V magnitude ~21).[20] This catalogue (Monet *et al.*, 2003) was produced from automated measuring-machine scans of photographic plates taken for the POSS (Palomar Observatory Sky Survey) and SERC-I, the UK Science and Engineering Research Council's first Southern Sky Survey. The proper motions are given in units of arcsec . y^{-1} with a precision of 0.002 arcsec . y^{-1}.

By way of comparison, the equatorial coordinates of HD73904 given in the USNO-B catalogue are ($08^h\,38^m\,23^s.91$; $-53°\,43'\,18''.39$) and its proper-motion coordinates (−26, 22) mas . y^{-1}, which are the equivalent in SI units of (2261.940, −937.622) mrad and (−126.05, 106.66) nrad . y^{-1}.

An example of a proper-motion catalogue derived from sequential observations over a period of three years, obtained primarily to search for stellar and exoplanetary microlensing, is that published by Sumi *et al.* (2004). Their catalogue, covering 49 selected fields, totalling approximately 0.003 4 sr (11 square degrees) of the galactic bulge surrounding the galactic centre, lists over 5 million stars with I-band monochromatic flux densities between ~100 mJy and ~150 μJy ($11 < I < 18$). Each entry in the catalogue includes an internal identity number, mean proper motion components in mas . y^{-1} for both equatorial and galactic coordinates, with standard deviations about the mean values, right ascension and declination (J2000.0) in decimal degrees, V magnitudes and (V − I) colours.

A partial entry from the proper-motion catalogue with angular measures as published and as converted to SI units is given in Table 5.6.

20 See http://www.nofs.navy.mil/data/fchpix/

5.6 The determination of the ages of celestial bodies

There are many different ways of determining the ages of celestial bodies, both in the universe and of the universe itself. In this chapter, only those methods that depend on the direct measurement of time or angular velocity will be considered, with examples of the use of radioactive decay (nucleocosmochronology) and the spin-down rate of pulsars.

5.6.1 Nucleocosmochronology

Nucleocosmochronology is defined by Schramm (1990) as '. . . the use of the abundance and production ratios of radioactive nucleides coupled with information on the chemical evolution of the Galaxy to obtain information about the time scales over which the solar system elements were formed'. The basic method for determining ages using this method was first set out by Rutherford (1929).

It is well known that some atomic nuclei are stable and some are not. Those that are not exist for a certain average length of time and then eject particles spontaneously (radioactivity). The new nucleus so formed, known as the daughter nucleus, may or may not be stable. For each radioactive nucleus there exists a transition probability λ, such that the nucleus will decay into its daughter nucleus within the next small interval of time $\mathrm{d}t$. The inverse of this transformation probability is the average lifetime of the parent nucleus and is measured in seconds (base SI unit) or Julian years. λ is measured in bequerels (a derived SI unit of dimension $[\boldsymbol{T}]^{-1}$ and symbol **Bq**), equal to the number of radioactive decays per second. A non-SI unit that is sometimes used to express the activity of a radioactive sample is the curie, equivalent to the mean rate of decay of 1 g of radium ($\sim 3.7 \times 10^{10}$ Bq).

Radioactivity has no 'memory', so it is not possible to predict when an individual nucleus will decay, but it is possible to make predictions about the average behaviour of large numbers of nuclei.

If a large number, N, of radioactive nuclei of a particular element exists at time t, then the number, $\mathrm{d}N$, that will decay in the next small time interval $\mathrm{d}t$, is given by:

$$\mathrm{d}N = -N\lambda\,\mathrm{d}t \tag{5.21}$$

or

$$\frac{\mathrm{d}N}{N} = -\lambda\,\mathrm{d}t$$

Integrating with respect to t from t_0 to $(t+t_0)$ yields the **law of exponential decay**:

$$N(t) = N(t_0)\,\mathrm{e}^{-\lambda t} \tag{5.22}$$

Figure 5.2 illustrates the shape of a typical radioactive decay curve.

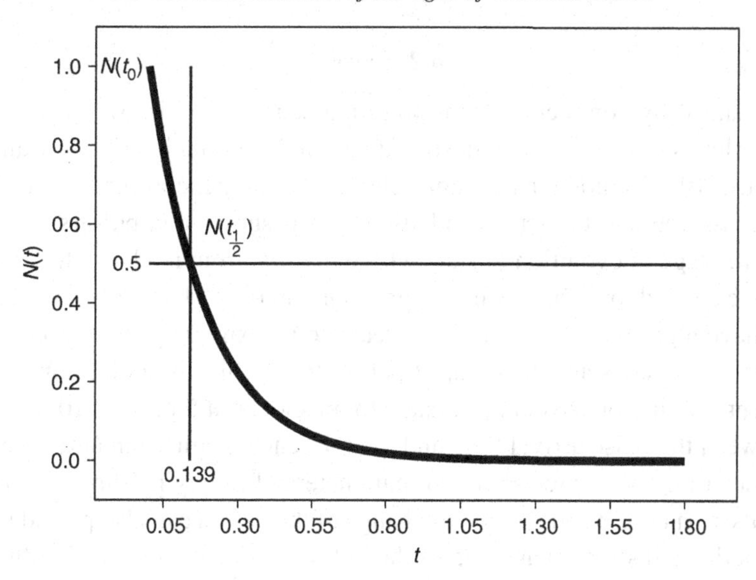

Figure 5.2. The radioactive decay of a large number of parent atomic nuclei into daughter atomic nuclei follows an exponentially shaped curve with time.

The **half-life**, $t_{\frac{1}{2}}$, of a radioactive nucleus is the time it takes, on average, for half of the original number of nuclei to decay into daughter nuclei:

$$t_{\frac{1}{2}} = \frac{\ln 2}{\lambda} \tag{5.23}$$

The **mean lifetime of a state** τ is defined by:

$$\tau = \frac{1}{\lambda} \tag{5.24}$$

After a time interval τ has elapsed, on average, approximately $(1 - 1/\mathrm{e})\%$ of the original nuclei have decayed into daughter nuclei. The half-life and the mean lifetime of the state are related by:

$$t_{\frac{1}{2}} = \tau \ln 2 \tag{5.25}$$

If at time $t = 0$, the number of parent nuclei is Np_0 and the number of daughter nuclei Nd_0 then, at time t, the number of daughter nuclei is:

$$\frac{Nd_t}{Np_t} = \mathrm{e}^{\lambda t} - 1 \tag{5.26}$$

From this equation, if λ is known and the ratio $Nd_t : Np_t$ can be measured observationally or experimentally, then a value can be determined for the age of the sample being investigated.

5.6.2 Pulsars

Pulsars, named by contraction from pulsating stars, were discovered in 1967 by Bell and Hewish using a newly designed and built radio telescope in Cambridge (Thorsett, 2001). Radio signals from pulsars take the general form of a series of extremely regular short bursts of radiation. It was shown that pulsars are neutron stars, a late stage of evolution of stars whose masses, from model stellar computations, are expected to lie between 0.2 and 2 solar masses ($4 \times 10^{29} - 4 \times 10^{30}$ kg). Pulsars have high rotational inertia that leads to an extremely stable rotation. The individual pulses are stable to around 1 part in 10^{14}, which is not far removed from the current stability of caesium fountain atomic clocks at 5 parts in 10^{16}. Comparison between the pulse arrival time and the reference atomic standard time can be used to determine very precise orbital parameters of binary pulsars. Observations of the pulsation period and the rate of change of the period can also provide data on the magnetic field strength and age of the pulsar (Burke & Graham-Smith, 2002). The slowing of the rotation period provides the energy for the pulses.

If the original period of pulsation at the time the pulsar was formed is P_0, the present pulsation period is P, and the rate of increase in the period is $\dot{P}$, then the current age of the pulsar, t, is:

$$t = \frac{P}{(n-1)\,\dot{P}}\left[1 - \left(\frac{P_0}{P}\right)^{n-1}\right] \tag{5.27}$$

If it is assumed that $P \gg P_0$ and $n = 3$ (for braking by magnetic spin-down radiation) then the characteristic age, τ, is defined as:

$$\tau = \frac{P}{2\,\dot{P}} \tag{5.28}$$

An actual value for n can only be directly determined if a value of the second derivative of P with respect to time can be measured.

The mean monthly signal-frequency data for the Crab pulsar used to plot Figure 5.3 covering the period from 1982 February 15 to 2002 September 15 was downloaded from the NVO (National Virtual Observatory of the USA) website.[21] The CRABTIME database was sourced from Lyne and collaborators at the Jodrell Bank Observatory,[22] with the addition of values for P and Ṗ (period of pulsation and its derivative with respect to time) calculated by HEASARC (the High Energy Astrophysics Science Archive Research Centre of NASA).[23]

[21] See http://nvo.stsci.edu/vor10/getRecord.aspx?id=ivo://nasa.heasarc/crabtime

[22] http://www.jb.man.ac.uk/pulsar/crab.html (provided by A.G. Lyne, C.A. Jordan & M.E. Roberts in *Jodrell Bank Crab Pulsar Timing Results, Monthly Ephemeris*).

[23] See http://heasarc.gsfc.nasa.gov/

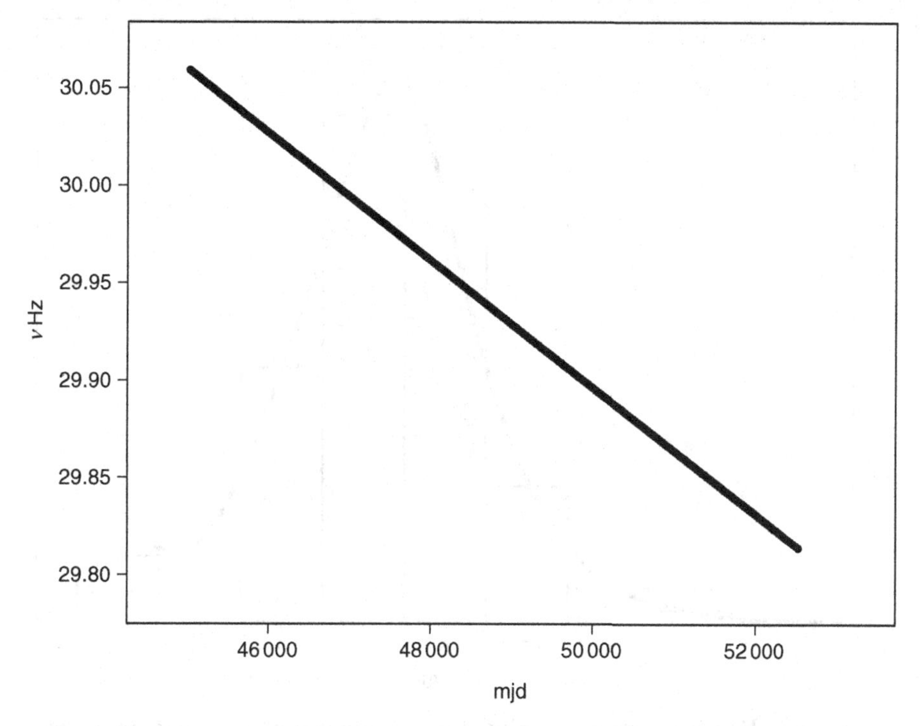

Figure 5.3. Crab pulsar rotation-rate slowdown from 1982 to 2002. The abscissa is given as modified julian day numbers and the ordinate as frequency in hertz.

Over the time covered in this data set the mean pulsation period, $\overline{P}$, was $0.033\,271 \pm 0.002\,106$ s and the mean rate of change of the pulsation period, $\overline{\dot{P}}$, was $4.192\,877 \times 10^{-13} \pm 0.265\,206 \times 10^{-13}$ s . s^{-1}. Substituting these values into Equation (5.28) and using an observational value of $n = 2.51 \pm 0.01$ for the braking index determined by Lyne *et al.* (1993), the age of the Crab pulsar is found to be 1665 y. This may be compared with the known age of 956 y at the present time (2010 CE). Burke & Graham-Smith (2002) warn that, at best, 'present day slow-down rates are only indications of actual age and not infallible measures of it'. One possible cause of this discrepancy, the effect of a large transverse motion on the observed period derivatives, was first pointed out by Shklovskii (1970). Applying a suitable correction for this (see Manchester *et al.*, 2005) produces a better value for the pulsar slow-down rate.

A catalogue of pulsar data compiled by Taylor *et al.* (1993) forms the basis of an online database, the ATNF Pulsar Catalog.[24] Some 1640 pulsars in the database have had their characteristic ages computed (using a braking index of 3), which

[24] See http://www.atnf.csiro.au/research/pulsar/psrcat/

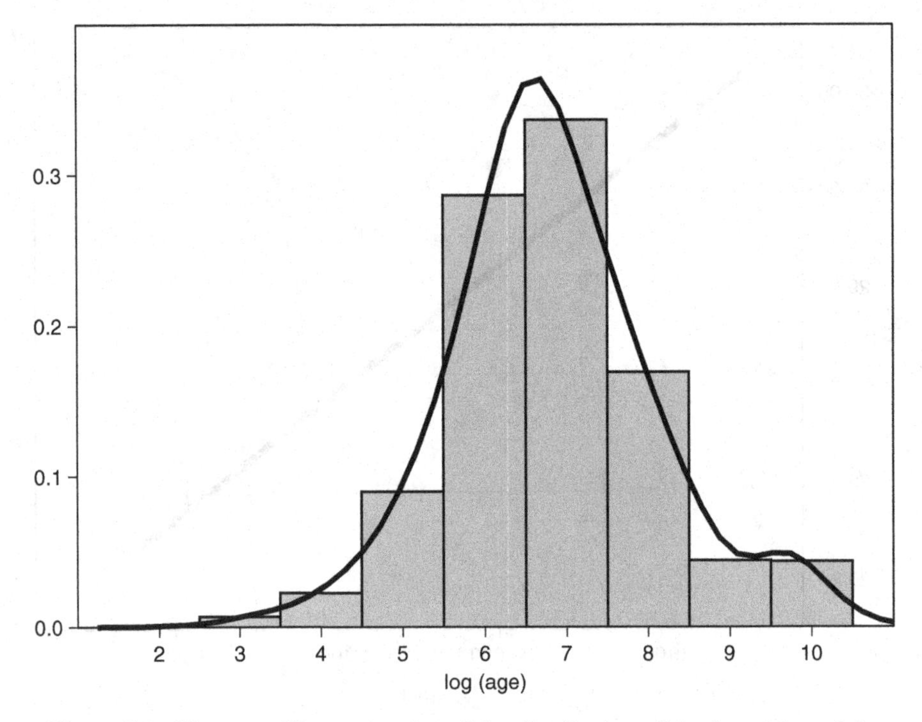

Figure 5.4. Histogram/frequency plot of the distribution of the logarithm of the characteristic ages of pulsars in Julian years listed in the ATFN Pulsar Catalog.

vary from the youngest at 218 y (PSR J1808-2024) to the oldest at 6.75×10^{10} y (PSR J0514-4002A)(see warning above!). An histogram of the logarithms of these pulsar ages is shown in Figure 5.4. The median age from the database is 5.47×10^6 y.

5.7 Summary and recommendations

5.7.1 Summary

Time was initially measured with reference to the rotation of the Earth and its period of revolution about the Sun (dynamical time). More recently, the development of very stable and accurate atomic clocks has led to the introduction of atomic time. The relationships between these two major time systems is given. Rotation periods and periods of revolution may be measured in derived SI units of rad . s^{-1}, some examples of which are given. A similar unit is used for the proper motions of the stars across the celestial sphere.

It is possible to determine the ages of some astronomical bodies by using the abundance and production ratios of radioactive isotopes. Decay rates from parent to daughter populations may be measured in bequerels, a unit of inverse time. The ages of pulsars may be found from a detailed study of the rate at which their rotation rate slows; an example of such a study is shown for the Crab pulsar.

5.7.2 Recommendations

The IAU recommends using the TAI system when precise measures of time intervals are required.

The commonly used unit for angular rotation, degrees per day, should be replaced by the SI unit, rad . s^{-1}. Similarly, stellar proper motion should use the same unit, though, given that its numerical value will be very small, rad . y^{-1} or submultiples thereof (e.g., mrad . y^{-1}, µrad . y^{-1} or nrad . y^{-1}) could be used. For very distant or slow moving bodies it is usual, at present, to quote the proper motion as angular measure per julian century.

The inverse unit of time can be either the hertz, when used as a derived SI unit of frequency, or the bequerel, when used as the number of radioactive decays per second. In the case of the latter, since the age of the celestial body is the unknown being sought, then the unit of interest becomes the inverse of the bequerel, which is the SI unit, the second.

6
Unit of length (metre)

6.1 SI definition of the metre

The **metre** is the length of the path travelled by light in vacuum during a time interval of 1/299 792 458 of a second.

The dimension of length is **[*L*]**, its unit is the **metre** and its symbol **m**.

The original definition of the metre was one ten millionth of the distance from the Earth's north pole to its equator, determined along a meridian arc that ran from Dunkirk in the north to Barcelona in the south. Observations were begun in 1792 by J. B. J. Delambre, who worked from Paris northwards and P. F. A. Méchain who made measurements from Paris to Barcelona. They completed their task in seven years and the metre thus determined was modelled in pure platinum as a one-metre-long bar (Alder, 2004).

6.2 Linear astronomical distances and diameters

The sizes of and the distances between astronomical bodies is generally extremely large by everyday terrestrial standards. This has resulted in astronomers inventing units such as the light year, the astronomical unit and the parsec, which are, at first sight, better able to deal with very large distances. The SI unit of length, the metre, used in conjunction with common prefixes is normally only used for measurements within the Solar System.

6.2.1 Size of the Earth

Were the Earth a perfect sphere it would follow from the original definition of the metre that its diameter would be $4\times10^7/\pi$ m. Unfortunately, however, the rotation of the Earth causes a flattening of the perfect sphere into an oblate spheroid, with the equatorial radius a being greater than the polar radius c. The amount of this flattening, f, or ellipticity ($= 1/f$), is given in McCarthy & Petit (2004)[25] as:

$$f = \frac{(a-c)}{a} = \frac{1}{298.25642} = 0.003\,352\,819\,70 \tag{6.1}$$

where $a = 6.378\,136\,60 \times 10^6\,\text{m}\,(\pm 0.10\,\text{m})$ and $c = 6.356\,751\,86 \times 10^6\,\text{m}$ (McCarthy & Petit, 2004).

A uniform reference spheroid, known as the **geoid**, is the imaginary surface that the Earth would have at mean sea level if an ellipse of semimajor axis a, and eccentricity e, were rotated about its minor axis (the polar axis in the case of the Earth). The eccentricity is derived from $e = \sqrt{(2f - f^2)}$, and the radius r, at geodetic latitude ϕ,[26] from:

$$r = a\,(1 - f\,.\sin^2\phi) \tag{6.2}$$

Example: derive the geocentric radius at the location of the Paris Observatory

The geodetic latitude of the Paris Observatory is (see *The Astronomical Almanac*, 1995).

$$\phi = 852.360629\,\text{mrad} \quad (\equiv 48^\circ\,50'.2)$$

where ϕ is measured northwards from the geodetic equator. Now substitute for ϕ, and for a and f (from McCarthy & Petit, 2004) in Equation (6.2) to determine the radius at mean sea level, r_{msl}:

$$r_{\text{msl}} = 6.3660165 \times 10^6\,\text{m}$$

To this value add the height of the Paris Observatory above mean sea level (= 67 m) to produce a final result for the geocentric radius at the location of the Paris Observatory of:

$$r_{\text{Paris Observatory}} = 6.366083\,5 \times 10^6\,\text{m}$$

The mean radius of the Earth

To obtain the mean radius, ρ, of the oblate spheroid, consider the volume of a sphere of radius ρ with the same volume as the oblate spheroid so that:

$$\frac{4}{3}\pi\,\rho^3 = \frac{4}{3}\pi\,(a^2 c) \tag{6.3}$$

[25] McCarthy & Petit cite the 1999 report of E. Groten to the IAG Special Commission SC3, Fundamental Constants, XXII IAG General Assembly.

[26] Woolard & Clemence (1966) define geodetic latitude as 'the angle between the geodetic vertical and the plane of the geodetic equator'.

or

$$\rho = (a^2 c)^{\frac{1}{3}} \tag{6.4}$$

Substituting the above values for a and c in Equation (6.4) produces a value for the mean radius of the geoid of $\rho = 6.371\,000\,37 \times 10^6$ m.

6.2.2 Distance to the Moon

The distance from the centre of the Earth to the Moon (see Figure 6.1) was initially determined, via trigonometry, from measurements of selected stars and the Moon, made at two observatories, one north and one south of the equator, which lay at approximately the same terrestrial longitude (e.g., the Royal Observatory, Greenwich, in London and the South African Astronomical Observatory in Cape Town).[27] The baseline, the straight line distance (i.e., through the Earth, not along the surface) between the two observatories, is computed knowing the latitude, longitude and height above mean sea level of each (now readily found from Global Positioning System (GPS) Earth satellite data).

This method has been superseded by using Earth-based laser ranging techniques. In the 1960s and 1970s, the Apollo astronauts placed optical retro-reflectors on the surface of the Moon. By transmitting short pulses of coherent light from a telescope on Earth to the Moon and recording the time of travel, t, of the beam from the Earth to the Moon and back, the distance, r, from the observer to the retro-reflector on the lunar surface is simply $r = ct/2$.

Given that the value of c is defined and that, by using atomic clocks, the value of t may be measured to better than 1 part in 10^{15}, this way of finding the distance to the Moon is far more accurate than the earlier trigonometric method. Recent measurements at the Apache Point Lunar Laser-ranging Operation have a median nightly range uncertainty of ± 1.8 mm (Battat *et al.*, 2009).

The mean distance from the centre of the Earth to the centre of the Moon is $3.844\,01 \times 10^8$ m $\pm$ 1000 m (Cox, 2000).

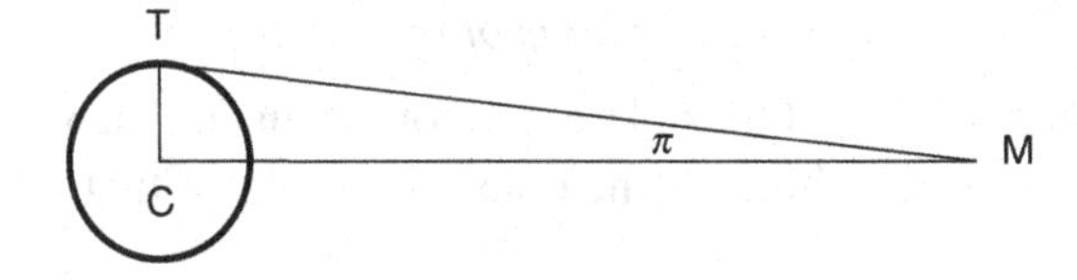

Figure 6.1. Horizontal parallax: point C is the centre of the Earth, angle MCT is a right angle, CT is the equatorial radius of the Earth, CM is the geocentric distance of the Moon (or Sun; for solar parallax, see Section 6.2.4) and π is the parallax angle.

[27] For details of the method, see Barlow & Bryan (1956).

The horizontal parallax of the Moon π_{Moon}, is the angle subtended by the equatorial radius of the Earth at the distance of the Moon. The mean value of the equatorial horizontal parallax is 16.593 271 8 mrad ($= 3422''.608$).

6.2.3 Astronomical unit

The definition of the astronomical unit in most textbooks on astronomy and astrophysics (e.g., Kutner, 2003; Burke & Graham-Smith, 2002; Lang, 2006 and even Cox, 2000) is generally something like: 'The astronomical unit is the mean distance from the Earth to the Sun and is equal to approximately 150 million kilometers or about 93 million miles.'

The conversion of the astronomical unit into SI units involves direct measures of the distances of planets and other Solar System bodies, which in the past relied on classical astrometric methods, such as trigonometric parallax, radial velocities and annual aberration with uncertainties of about 1 part in 10^4. The value of the gravitational constant, $G\,(=k^2)$ (see Equation 6.8 below), is also only known to about 1 part in 10^5. These have led to computations in celestial mechanics being carried out using the IAU set of astronomical units: the day as the unit of time; the mass of the Sun as the unit of mass; and the astronomical unit as the unit of length. This has allowed a far greater accuracy in the preparation of planetary ephemerides. However, this situation is changing with the far higher accuracy of Earth-based radar-determined planetary distances aided by atomic clocks and the use of space probes (see, e.g., Muhleman *et al.*, 1962, and Pitjeva, 2005).

Definition of the astronomical unit, version 1, Kepler's third law

In 1618, Kepler published what became known as his third law of planetary motion, which states that the squares of the sidereal periods of revolution of any two planets about the Sun are proportional to the cubes of their mean distances from the Sun, i.e.,

$$\frac{P_1^2}{P_2^2} = \frac{a_1^3}{a_2^3} \tag{6.5}$$

If one of the planets is the Earth, with a period of revolution about the Sun of one sidereal year and a mean solar distance of one astronomical unit, then Equation (6.5) may be rewritten as:

$$\frac{P^2}{a^3} = 1 \tag{6.6}$$

where P is the period of revolution about the Sun of a gravitationally bound astronomical body other than the Earth and a is its mean distance from the Sun in astronomical units.

For any other planet with sidereal period P years Equation (6.5) may be rewritten as:

$$a = P^{\frac{2}{3}} \tag{6.7}$$

Newton expressed Kepler's third law in a more general form, which involved the constant of gravitation k, the mass of the planet m, and the mass of the Sun $M_\odot$:

$$a = \left[\frac{k^2\, P^2\,(M_\odot + m)}{4\pi^2}\right]^{\frac{1}{3}} \tag{6.8}$$

or when expressed in IAU astronomical units (see below, version 3):

$$a = \left[\frac{k^2\, P^2\,(1 + m)}{4\pi^2}\right]^{\frac{1}{3}} \tag{6.9}$$

Definition of the astronomical unit, version 2, IAU(1976)

In 1976, the IAU[28] redefined the astronomical unit to be equal to the distance from the centre of the Sun at which a particle of negligible mass, in an unperturbed circular orbit, would have an orbital period of 365.256 898 3 days where 1 day is defined to be 86 400 SI seconds.

Definition of the astronomical unit, version 3, AAO(2010)

The Astronomical Almanac Online[29] provides the following formal definition of the astronomical unit of length:

The astronomical unit of length is that length (A) for which the Gaussian gravitational constant (k) takes the value 0.017 202 098 95 when the units of measurement are the astronomical units of length, mass and time. The dimensions of k^2 are those of the constant of gravitation (G).

That is, $A^3 . S^{-1} . D^{-2}$, $(\mathrm{dim}_{\mathrm{IAU}}([A]^3 . [S]^{-1} . [D]^{-2}) \equiv \mathrm{dim}_{\mathrm{SI}}([L]^3 . [M]^{-1} . [T]^{-2}))$, where S is the astronomical unit of mass equal to the mass of the Sun and D is the astronomical unit of time equal to one day of 86 400 SI seconds.

Definition of the astronomical unit, future versions?

All the versions of the definition of the astronomical unit given above are based on Newtonian dynamics. So an obvious development in redefining the astronomical unit is to use the framework of general relativity. Examples of this approach are given by, e.g., Huang *et al.* (1995), Klioner (2007) and in a summary by Capitaine & Guinot (2008).

[28] See http://www.iau.org/public-press/themes/measuring/
[29] See http://asa.usno.mil/index.html

Also, given the precision with which the astronomical unit can now be defined, the assumption that the astronomical unit of mass (one solar mass) is a constant is no longer acceptable. It is known that the Sun converts mass into energy resulting in a loss of mass equal to 4.3×10^9 kg . s^{-1}, as well as further particulate mass loss through the solar wind of approximately 1.3×10^9 kg . s^{-1}. A total of some 5.5 million tonnes per second, or, in IAU astronomical unit terms: $2.405 \times 10^{-16}\, M_\odot \,.\, \mathrm{d}^{-1}$, or in units of solar masses per century: $8.797 \times 10^{-12}\, M_\odot$. century^{-1}. According to Noerdlinger (2008), such a loss will result in the orbits of the planets expanding at the same relative rate and their periods of revolution to increase at twice that relative rate, causing the planet Mercury to drift away from its predicted position by more than 5 km in an interval of 200 years.

Another way in which the astronomical unit may be increased is via the total conservation of angular momentum law, proposed by Miura *et al.* (2009).

It is apparent that further revision of the definition of the astronomical unit of length is not only desirable but necessary. However, given that such revisions may take many years to carry out, the only presently available solution to convert astronomical units to metres is to use the currently accepted best conversion value.

Recent determinations of the astronomical unit

By combining more than 300 000 positional observations, both optical (transit circle, photographic and CCD) and radio, of the planets and spacecraft made from 1913 to 2003, Pitjeva (2005) determined the astronomical unit to be: $1.495\,978\,706\,960 \times 10^{11}$ m ± 0.1 m.

This value was updated by Pitjeva & Standish (2009) using improved ephemeris computational methods at JPL in Pasadena and at IAA RAS in St Petersburg to: $1.495\,978\,707\,00 \times 10^{11}$ m ± 0.3 m.

McCarthy & Petit (2004) in the *IERS Conventions (2003)* list a scale factor (km/au) of 149 597 870.691 as an auxiliary constant from the JPL Planetary and Lunar Ephemerides (DE405/LE405). In metres, this conversion factor is: 1 au $= 1.495\,978\,706\,91 \times 10^{11}$ m (TDB) or **1 au = $\mathbf{1.495\,978\,714\,64 \times 10^{11}}$ m (SI)**. This is the scaling factor that will be used in subsequent computations.

Table 6.1 gives the distances to all the planets and some of the dwarf planets in both astronomical units and the SI unit of length, the metre. As can be seen from columns 3 and 4 in Table 6.1, distances within the Solar System may be readily expressed in either IAU astronomical units or prefixed values of the SI metre.

6.2.4 Solar parallax

The Sun's equatorial horizontal parallax, $\pi_\odot$, was defined as the angle at the centre of the Sun subtended by the Earth's equatorial radius, $a_\oplus$, at a distance of 1

Table 6.1. *Sidereal period in days and distances in IAU astronomical units and SI metres for the planets and some dwarf planets in the Solar System*

Planet	Sidereal period IAU d	Semimajor axis astronomical units	Semimajor axis $\times 10^{12}$ m = Tm
Mercury	87.968	0.387 098 93	0.057 909 17
Venus	224.695	0.723 331 99	0.108 208 92
Earth	365.242	1.000 000 11	0.149 597 88
Mars	686.930	1.523 662 31	0.279 366 38
Jupiter	4330.596	5.203 363 01	0.778 412 03
Saturn	10 746.940	9.537 070 32	1.426 725 42
Uranus	30 588.740	19.191 263 93	2.870 972 23
Neptune	59 799.900	30.068 963 48	4.498 252 93
Dwarf Planet			
Ceres	1 680.150	2.767 000 00	0.413 937 31
Pluto	90 656.073	39.481 686 77	5.906 376 30
Eris	204 865.073	67.959 000 00	10.166 521 70

astronomical unit, when the astronomical unit itself was defined to be the mean Earth–Sun distance:

$$\pi_{\odot} = \frac{a_{\oplus}}{1\,\text{au}} \tag{6.10}$$

From Section 6.2.1, the equatorial radius of the Earth is $6.378\,136\,60 \times 10^6$ m and from the previous section the value of the astronomical unit in metres is given, so substituting these numerical values into Equation (6.10) gives the solar parallax to be: 42.635 209 56 μrad (equivalent to 8″.794 143 240). As the Earth's orbit is elliptical, the value of the solar parallax varies from approximately 43.342 μrad (8″.94) at perihelion on January 4, when the distance to the Sun is about 0.983 au, to 41.936 μrad (8″.65) at aphelion on July 4, when the distance to the Sun is about 1.017 au.

6.2.5 Horizontal parallax via gravitational microlensing

Generally, the horizontal parallax of celestial bodies beyond the Solar System is so small that it would not be measurable from the Earth. However, Gould *et al.* (2009), from observations made by several of the worldwide microlensing groups, have been able to measure the effects of terrestrial parallax. The object observed, OGLE-2007-BLG-224, was shown to be a very low mass thick-disc brown dwarf at a distance of 525 ± 40 pc. The equivalent parallax, π_{orbit}, in SI units is 9.235 nrad. The angle $\pi_{\oplus}$, subtended by the semimajor axis, $a_{\oplus}$, of the Earth at the distance of

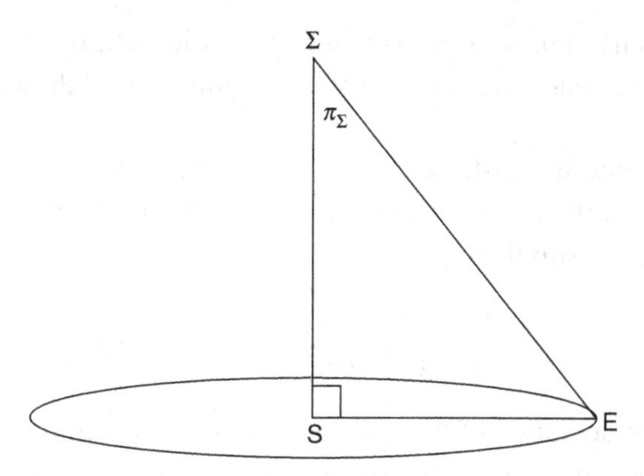

Figure 6.2. Trigonometric parallax: point S is the Sun, SE is the mean radius of the Earth's orbit about the Sun, SΣ is the heliocentric distance of the celestial body Σ and π_Σ is the parallax angle of Σ. NB: this figure is NOT to scale since the distance SΣ ≫ SE.

the microlensing brown dwarf is approximately given by:

$$\pi_\oplus = \pi_{\text{orbit}} \cdot \frac{a_\oplus}{1\,\text{au}} = 3.937\,168 \times 10^{-13}\,\text{rad} \tag{6.11}$$

or 394 frad (femtoradians).

6.2.6 Trigonometric parallax

The trigonometric or heliocentric parallax of a celestial body is the angle subtended by the radius of the Earth's orbit about the Sun at the distance of the celestial body (see Figure 6.2).

Given that the relative size of the measurement errors in determining the parallax angle is large in comparison with that of determining the Earth–Sun distance, it may be assumed that the Earth–Sun distance is a good approximation to 1 astronomical unit. Weygand *et al.* (1999) suggest, taking into account the gravitational influences of the major planets, that the Earth's orbit may best be described as an ellipse, with semimajor axis equal to 1.000 000 2 astronomical units. So the difference between the measured value of the Earth–Sun distance and that determined from version 3 of the astronomical unit definition amounts to 2 parts in 10^7. Perryman *et al.* (1997) give the median value of the standard deviation of the mean parallax for all stars measured with a monochromatic flux density greater than approximately 1 Jy ($H_p \leq 9$) as ±4.7 nrad (≡ 0.97 mas). The entry in the HIPPARCOS catalogue[30] for the bright nearby star α Cen A is $\pi = 3597.9 \pm 6.8$ nrad (≡ 742.12 ± 1.40 mas). The standard deviation of the measurements of the parallax to the value of the

[30] See http://archive.eso.org/skycat/servers/ASTROM

parallax is the equivalent of approximately 1 part in 530, or some 20 000 times greater than the differences due to the differing definitions of the astronomical unit.

Example: determine the distance to α Cen A in metres
Given the mean parallax π of α Cen A as 3 597.9 nrad, then its distance, D, in astronomical units, is simply:

$$D = \frac{A}{\pi} = \frac{1}{3\,597.9 \times 10^{-9}} = 277\,940 \tag{6.12}$$

Using the standard deviation of the parallax distance range limits, $D_{\max}$ and $D_{\min}$, for the star may be calculated, where:

$$D_{\max} = \frac{1}{\pi - \sigma_\pi}$$
$$D_{\min} = \frac{1}{\pi + \sigma_\pi} \tag{6.13}$$

Substituting the catalogue values for the mean parallax and its standard deviation, the distance in astronomical units to α Cen A may be given as: $277\,940^{278\,465}_{277\,416}$. To convert these distances into metres, the SI unit of length, simply multiply by the final conversion factor given in Section 6.2.3, which gives: 4.158×10^{16} m as the mean distance within the limits 4.150×10^{16} m to 4.166×10^{16} m. The mean distance may also be expressed as 41.579 Pm (petametres) or 4.158 d 16 m.

In Table 6.2, parallaxes in mas and nrad, and distances in astronomical units, SI metres, parsecs and light years derived from the SIMBAD[31] database are given. For this group of stars, which relative to the size of the Milky Way galaxy may be considered as very close, the distance measurements in astronomical units become unwieldy, and the name of the unit does not lend itself to the addition of prefixes. Astronomers have accordingly invented two larger measurements of distance: the parsec and the light year.

6.2.7 Parsec

The **parsec** is the distance, D, at which one astronomical unit subtends an angle of one arc second. It is the reciprocal of the parallax π, expressed in arcseconds of a celestial body, i.e.:

$$D = \frac{1}{\pi} \tag{6.14}$$

[31] See http://simbad.u-strasbg.fr/simbad/sim-fid, which refers to Perryman *et al.* (1997).

Table 6.2. *Parallaxes and distances in astronomical units, metres, parsecs and light years for the ten brightest stars in the night sky*

Star name	Parallax π_{mas} π_{nrad}	Distance astronomical units	Distance (SI) metres $\times 10^{15}$	Distance parsecs light years
Sirius	379.21 1838.462	5.439×10^5	81.371	2.637 8.601
Canopus	10.43 50.566	1.978×10^7	2958.464	95.877 312.710
α Cen A	742.24 3598.481	2.779×10^5	41.572	1.347 4.394
Arcturus	88.85 430.757	2.321×10^6	347.291	11.255 36.709
Vega	128.93 625.070	1.600×10^6	239.330	7.756 25.297
Capella	77.29 374.712	2.669×10^6	399.234	12.938 42.199
Rigel	4.22 20.459	4.888×10^7	7312.032	236.967 772.882
Procyon	285.93 1386.228	7.214×10^5	107.917	3.497 11.407
Betelgeuse	7.63 36.991	2.703×10^7	4044.138	131.062 427.466
Achernar	22.68 109.956	9.095×10^6	1360.528	44.092 143.808

If the parallax is expressed in radians π_{rad}, then Equation (6.14) may be rewritten as:

$$D = \frac{1}{206\,264.806\,265 \times \pi_{\text{rad}}} \tag{6.15}$$

The dimension of the parsec, being the reciprocal of an angular measure, is $[L]\,.\,[L]^{-1}$ and its symbol is **pc**.

Example: determine the distance to the star Achernar

In Table 6.2, Achernar is listed as having a parallax of 109.956 nrad, hence from Equation (6.15):

$$\frac{1}{109.956 \times 10^{-9} \times 206264.806265} = 44.092\,\text{pc}$$

For stars in the solar neighbourhood the parsec is a useful unit. When considering the Galaxy as a whole, the **kiloparsec** equal to 1000 pc is preferable. For extragalactic distances, the **megaparsec**, equal to 1 000 000 pc, is commonly used with, for very large distances, the **gigaparsec**, equal to 1 000 000 000 pc.

In SI units, the parsec is equal to, **3.085 677 6** $\times$ **10**^{16} **m**, and in IAU units is equal to **206 264.806 265 au**.

The parsec is calculated directly from a measurement of the parallax of an astronomical object.

6.2.8 Light year

The **light year** may be considered to be a derived, defined unit equal to the distance that light, travelling at 299 792 458 m s^{-1}, covers in 1 julian year of exactly 3.15576×10^{7} (SI) s. Hence, 1 light year is exactly equivalent to **9.460 730 472 580 8** $\times$ **10**^{15} **m**.

Its dimension is $[L].[T]^{-1}.[T]=[L]$ and its symbol is **ly**.

To convert parsecs into light years, multiply by **3.261 563 8**, and to convert astronomical units into light years, divide by **63 241.077 101 522**.

The distance in light years is not directly measurable for even the nearest star, though such measurements may become possible in the far distant future via radar or laser reflections from bodies orbiting nearby stars, or from signals received from probes sent to such bodies.

At present, the parsec is the unit of choice for professional astronomers and the writers of university texts, whilst the light year is the preference for authors of popular astronomical articles and books. Note that the light year is not even a non-SI unit recognized for use in astronomy, neither is it a unit recommended by the IAU (see Wilkins, 1989). A brief history of the development of the parsec and the light year is given by Beech (2008), in which he points out that the first indirect reference (Bessel, 1838) to the term 'light year' predates the published appearance of the parsec (Dyson, 1913) by some 75 years.

6.2.9 Some examples of astronomical distances

In Table 6.3, the distance in metres is given both in decimal notation (see Section 2.5.5) and using SI prefixes (Em = Exametres, Zm = Zettametres and Ym = Yottametres). The distances in parsecs listed in the table were extracted from WEBDA,[32] the NASA/IPAC Extragalactic Database,[33] Cox (2000), *The Astronomical Almanac* (1995) and van Leeuwen (2007). Note that the distances given for M104, the Coma cluster of galaxies and the A3266 cluster are galactocentric rather than heliocentric. A number density plot of the cluster A3266 is given in Figure 4.10.

[32] http://www.univie.ac.at/webda/
[33] http://nedwww.ipac.caltech.edu/

Table 6.3. *Some examples of distances given in metres, parsecs and light years to astronomical bodies lying outside the Solar System*

Object name	Object type	Distance (SI) metres	Distance parsecs, kpc or Mpc	Distance light years
Hyades	galactic star cluster	1.389 d 18 1.389 Em	45 pc	147
Pleiades	galactic star cluster	3.765 d 18 3.756 Em	122 pc	398
IC2391	galactic star cluster	4.536 d 18 4.536 Em	147 pc	479
M67	galactic star cluster	2.802 d 19 28.02 Em	908 pc	2961
NGC6572 (Ring Nebula)	planetary nebula	2.006 d 19 20.06 Em	650 pc	2120
47 Tuc	globular star cluster	1.419 d 20 141.9 Em	4.6 kpc	15 003 1.500 d 4
M3	globular star cluster	3.086 d 20 308.6 Em	10 kpc	32 616 3.262 d 4
LMC	local group galaxy	1.697 d 21 1.697 Zm	55 kpc	179 386 1.793 d 5
M31	local group galaxy	2.237 d 22 22.37 Zm	725 kpc	2 364 634 2.364 d 6
M104 (Sombrero Hat)	edge on galaxy	6.171 d 23 617.1 Zm	20 Mpc	65 231 276 6.523 d 7
Coma	cluster of galaxies	2.934 d 24 2.934 Ym	95 Mpc	310 174 717 3.102 d 8
A3266 (Abell cluster)	cluster of galaxies	7.387 d 24 7.387 Ym	239 Mpc	780 818 374 7.808 d 8

In this table the shorthand notation $m\,\mathrm{d}\,n$ has been used where $m\,\mathrm{d}\,n \equiv m \times 10^n$.

The simple prefixes work well for the parsec unit and are easy to remember, and those for the metre also work well, though are less easy to remember and may be difficult to associate immediately with the relevant power of 10. The d notation does not have this problem. Prefixes are not commonly used with light years, but with numbers of light years not exceeding tens of billions this is not an insurmountable difficulty (e.g., the Coma cluster of galaxies would be said to be at a galactocentric distance of approximately 310 million light years).

6.3 Linear motion

The **linear velocity** of a moving body is its rate of change in position with time in a particular direction.

The dimension of the SI derived unit of linear velocity is $[L].[T]^{-1}$ its name is the **metre per second** and its symbol is $\mathbf{m\,.\,s^{-1}}$.

6.3.1 The speed of electromagnetic radiation

In 1975, the 17th CPGM recommended that the speed of electromagnetic radiation in vacuo, c, be a physical constant, defined to be exactly 299 792 458 m . s^{-1}.

A fixed speed of electromagnetic radiation not only relates distance in metres and time in SI seconds but also frequency ν, in hertz, and wavelength λ, in metres, via:

$$\dim(c) = [L].[T]^{-1} \tag{6.16}$$

and

$$c = \lambda . \nu \tag{6.17}$$

Since it is possible to measure time more accurately than any other SI unit, the above relationships lead to a greatly improved realization of the metre that is limited solely by the accuracy of the time reference.

The value selected for the defined c is very close to that determined by Evenson *et al.* (1972) at the US National Institute of Standards and Technology (NIST) in Boulder, Colorado, in 1972. The experimentally measured velocity was 299 792 456.2 $\pm$ 1.1 m . s^{-1}.

Figure 6.3 illustrates the relationship between frequency and wavelength throughout the electromagnetic spectrum, from short wavelength, high-energy and high-frequency γ-rays to long wavelength, low-energy and low-frequency microwave and radio waves.

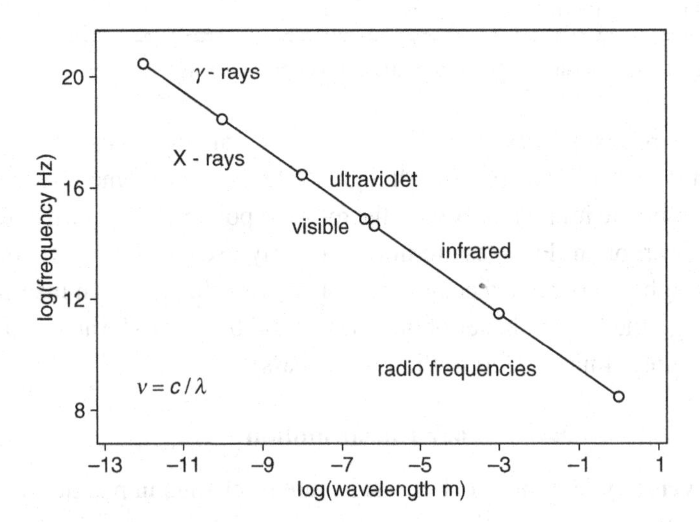

Figure 6.3. A plot showing the relationship between log(frequency) measured in Hz and log(wavelength) measured in m. Named portions of the electromagnetic spectrum are shown.

6.3.2 Radial velocity

The radial velocity of an astronomical body is the component of its motion along the line of sight from the observer to the body. The radial velocity component is at right angles to the transverse component across the celestial sphere.

It is determined by measuring the shift of spectral lines in the electromagnetic radiation emitted by the celestial body when compared with a laboratory set of spectral lines, which are at rest relative the observer. If the wavelength of an identified spectral feature in the spectrum of the astronomical body is λ, and the wavelength measured in the laboratory is λ_0, then the radial velocity v, of the astronomical body where $v \ll c$, is given by:

$$v = c\,\frac{(\lambda - \lambda_0)}{\lambda_0} = c\,\frac{(\nu_0 - \nu)}{\nu} \tag{6.18}$$

where c is the speed of electromagnetic radiation. Radial velocity may also be defined in terms of observed frequency ν, and laboratory frequency ν_0. For bodies moving towards the observer, the spectral features (lines or groups of lines or bands) are shifted towards shorter wavelengths (higher frequencies) and are said to be 'blue shifted' with a negative radial velocity, whilst those bodies moving away have a positive radial velocity and their spectral features are said to be 'red shifted'. A positive radial velocity corresponds to lower measured frequencies, whilst negative radial velocities correspond to higher measured frequencies.

Table 6.4 lists the heliocentric radial velocities of the ten brightest stars in the night sky, with an associated measurement error, a quality rating from A = best quality measurement to E = worst quality measurement and a simple description of the star. This information was extracted from the SIMBAD astronomical database.[34] Note that all the radial velocity values are less than $100\,\mathrm{km\,.\,s^{-1}}$, implying that the $\mathrm{km\,.\,s^{-1}}$ is a sensible SI unit to use for radial velocities. In fact, since the largest radial velocity that could be measured is less than $300\,000\,\mathrm{km\,.\,s^{-1}}$ and the smallest that is currently being measured is approximately $1\,\mathrm{m\,.\,s^{-1}}$,[35] all radial velocities are comfortably covered by the SI unit for linear velocity.

6.3.3 Space motion of stars

The components of a star's space velocity, v, are its radial velocity v_r along the line of sight to the star, and its transverse velocity v_t at right angles to that line of sight. If the motion of the star makes an angle θ to the line of sight, then:

$$v_r = v\cos\theta$$

[34] See http://simbad.u-strasbg.fr/simbad/sim-fid

[35] In radial velocity searches for exoplanets (planets which orbit stars other than the Sun).

Table 6.4. *Radial velocities and brief classifications for the ten brightest stars in the night sky*

Star name	Radial velocity km . s^{-1}	Quality rating	SIMBAD description
Sirius A	-7.6 ± 0.9	A	spectroscopic binary
Canopus	$+20.5 \pm 0.9$	A	star
α Cen A	-24.6 ± 0.9	A	double star
Arcturus	-5.2 ± 0.9	A	variable star
Vega	-13.9 ± 0.9	A	variable star
Capella	$+30.2 \pm 0.9$	A	RS Cvn type variable star
Rigel	$+20.7 \pm 0.9$	A	emission-line star
Procyon	-3.2 ± 0.9	A	spectroscopic binary
Betelgeuse	$+21.91 \pm 0.51$	B	semiregular pulsating star
Achernar	$+16 \pm 5$	C	Be star

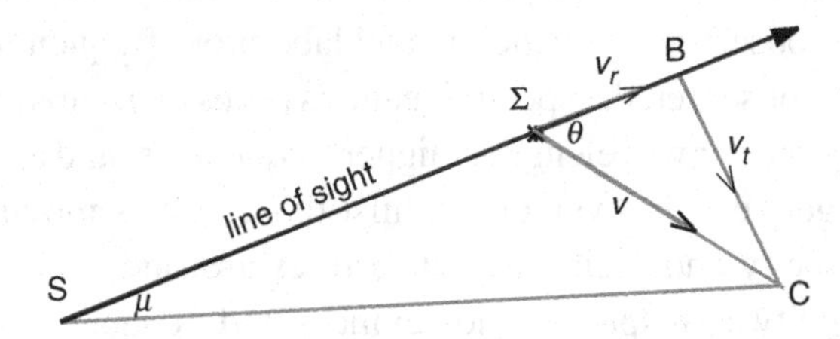

Figure 6.4. A plot showing the relationship between the transverse and radial velocity components and the space velocity of a star. The proper-motion angle is labelled μ, the Sun S, the star Σ, and the line of sight SΣB. The angle ΣBC is a right angle.

and

$$v_t = v \sin \theta \tag{6.19}$$

In SI units the radial velocity is measured in **m . s**$^{-1}$. The transverse component is the product of the distance r, in metres to the star, and its angular proper motion μ, in **rad . s**$^{-1}$, so:

$$v_t = r \,.\, \mu \tag{6.20}$$

Radial and transverse components are combined to yield the space velocity v of the star, thus:

$$v = \sqrt{(v_t^2 + v_r^2)} \tag{6.21}$$

The various components are shown in Figure 6.4.

Example: determine the space velocity of Vega

Consider the prominent nearby A-type star Vega. From Table 6.2 its distance is 2.3933×10^{17} m, from Table 6.4 its radial velocity is $-13\,900\,\mathrm{m\,.\,s^{-1}}$ and from Table 5.4 its equatorial coordinate proper-motion components are $(3.088\times10^{-14}, 4.416\times10^{-14})\,\mathrm{rad\,.\,s^{-1}}$. Combine the proper motion components to give $\mu = 5.389\times10^{-14}\,\mathrm{rad\,.\,s^{-1}}$. Now multiply μ by the distance to Vega in metres to obtain the transverse velocity in $\mathrm{m\,.\,s^{-1}}$, $v_t = 12\,900$. Finally, take the square root of the sum of the squares of the velocity components:

$$v = \sqrt{(-13\,900)^2 + (12\,900)^2} = \pm\ 18\,961\,\mathrm{m\,.\,s^{-1}} \tag{6.22}$$

or, using an appropriate prefix, $v = \pm\,18.96\,\mathrm{km\,.\,s^{-1}}$.

Since, from its measured radial velocity, it is known that the star Vega is moving towards the Sun, the correct space velocity is the negative solution, so $v = -18.96\,\mathrm{km\,.\,s^{-1}}$.

6.3.4 Red shifts and look-back times

The definition of radial velocity given in Equation (6.17) is strictly Newtonian and applies only when $v \ll c$. Extragalactic objects that lie beyond the local group of galaxies have spectra which are shifted towards the red with respect to a comparison terrestrially based laboratory spectrum. The amount of the shift z is accordingly known as the **red shift** and is defined, in both wavelength and frequency terms, as:

$$z = \frac{\lambda - \lambda_0}{\lambda_0} = \frac{\nu_0 - \nu}{\nu} \tag{6.23}$$

or, by a simple rearrangement:

$$z = \frac{\lambda}{\lambda_0} - 1 = \frac{\nu_0}{\nu} - 1 \tag{6.24}$$

Being a ratio of wavelengths or frequencies, the red shift is of dimension 1 and hence is a dimensionless unit:

$$\dim(z) = [L]\,.\,[L]^{-1} = [T^{-1}]\,.\,[T^{-1}]^{-1} = 1 \tag{6.25}$$

For radial velocities, $v \ll c$, the relationship between z and v may be written as:

$$v = c\,.\,z \tag{6.26}$$

Hubble (1929) showed that the velocity v, with which a galaxy is receding (its **recession velocity**), is directly proportional to its distance D, i.e.:

$$v = H_0\,.\,D = c\,.\,z \tag{6.27}$$

For very distant galaxies, the recession velocity is no longer very small in comparison with the velocity of light and the relationship has to be written in the relativistic form:

$$v = c \, . \left[\frac{(z+1)^2 - 1}{(z+1)^2 + 1} \right] \tag{6.28}$$

There are three component parts to a red shift: that due to the actual or peculiar motion of the celestial object itself; that due to the gravitational field of the object and known as the gravitational red shift; and that due to the expansion of the Universe and known as the cosmological red shift.

Since, in general, celestial objects are detected by the electromagnetic radiation they emit, and given that the speed of such radiation is finite, then the further away an object appears to be, the longer ago the detected radiation was emitted. This leads to yet another unit in common use the **look-back time** or sometimes its complement, the **time elapsed since the Big Bang**, which occurred some 13.7 billion years ago. Note that the sum of these two numbers is the age of the Universe at the time the observation was made. The dimension of look-back time and time elapse since the Big Bang is $[\boldsymbol{T}]$ and its SI unit is the second and its IAU unit the Julian year.

Wright (2006) has constructed an online calculator[36] which, for a given cosmological model and input value of red shift, computes distances in parsecs and light years, light travel times, the age of the Universe at the departure time of the electromagnetic radiation and other variables of use in cosmological studies.

6.4 Acceleration

The dimension of the SI derived unit of acceleration is $[\boldsymbol{L}]\,.\,[\boldsymbol{T}]^{-2}$, its unit is the **metre per second per second** and its symbol is $\mathbf{m\,.\,s^{-2}}$.

The acceleration at the surface of a celestial body due to its gravity is known as its **surface gravity**.

In Cox (2000), the values of the surface gravity of the bodies in Table 6.5 are given in cgs units. To convert to SI units, divide the cgs value by 100, e.g., the surface gravity of Mars:

$$\text{Mars}_{(\text{SI})} = \frac{\text{Mars}_{(\text{cgs})}}{100} = \frac{371}{100} = 3.71\,\text{m}\,.\,\text{s}^{-2}$$

Whilst cgs units are still in common use for surface gravity measurements, SI units are now readily accepted and recommended by the IAU.

36 See http://astro.ucla.edu/~wright/DlttCalc.html

Table 6.5. *Surface gravities of some bodies in the Solar System*

Name	Description	Surface gravity m . s^{-2}	Name	Description	Surface gravity m . s^{-2}
Mercury	planet	3.70	Pluto	dwarf planet	0.81
Venus	planet	8.87	Moon	satellite	1.62
Earth	planet	9.81	Sun	star	274.0
Mars	planet	3.71			
Jupiter	planet	23.12			
Saturn	planet	8.96			
Uranus	planet	8.69			
Neptune	planet	11.0			

6.5 Area

The dimension of the SI unit of area is $[L] . [L] = [L]^2$, its name is the **square metre** and its symbol is $\mathbf{m}^2$.

This unit is in common use in astronomy, as are the prefixed units cm^2 and km^2, and surface areas of planets relative to that of the Earth and the stars relative to that of the Sun.

Celestial bodies such as the planets and stars generally have a regular shape and are mainly either spheres or oblate spheroids.[37] The surface area, A of such bodies, with an equatorial radius of a, a polar radius of c, and values of oblateness[38] $\omega \left(= \frac{a-c}{c}\right)$ and eccentricity $e \left(= \frac{\sqrt{a-c}}{c}\right)$ are given by:

$$A = \left\| \begin{array}{ll} 4\pi a^2 & \text{if} \quad c = a \\ 2\pi a^2 + \frac{\pi c^2}{e} . \ln\left[\frac{1+e}{1-e}\right] & \text{if} \quad a > c \end{array} \right. \tag{6.29}$$

6.6 Volume

The dimension of the SI unit of volume is $[L] . [L] . [L] = [L]^3$, its name is the **cubic metre**, and its symbol is $\mathbf{m}^3$.

This unit is in common use in astronomy, as are the prefixed SI units cm^3 and km^3, and volumes of planets relative to that of the Earth and the stars relative to that of the Sun.

[37] See http://mathworld.wolfram.com/OblateSpheroid.html
[38] Note that the mathematical expressions for oblateness and flattening are identical.

Table 6.6. *Radii, surface areas and volumes of planets and one dwarf planet in the Solar System*

Name	Equatorial radius m km	$r_{\oplus}$	Surface area m^2	$A_{\oplus}$	Volume m^3	$V_{\oplus}$
Mercury	2.43976 d 6 2 439.76	0.383	7.480 d 13	0.147	6.083 d 19	0.056
Venus	6.052 30 d 6 6 052.30	0.9488	4.603 d 14	0.902	9.286 d 20	0.857
Earth	6.378 14 d 6 6 378.14	1.0000	5.101 d 14	1.000	1.083 d 21	1.000
Mars	3.397 52 d 6 3 397.52	0.533	1.444 d 14	0.283	1.632 d 20	0.151
Jupiter	7.149 2 d 7 71 492	11.209	6.164 d 16	120.838	1.437 d 24	1327
Saturn	6.026 8 d 7 60 286	9.449	4.295 d 16	84.213	8.351 d 23	771
Uranus	2.555 9 d 7 25 559	4.007	8.087 d 15	15.854	6.837 d 22	63.1
Neptune	2.476 4 d 7 24 764	3.883	7.620 d 15	14.939	6.254 d 22	57.739
Pluto	1.195 d 6 1 195	0.187	1.795 d 13	0.035	7.148 d 18	0.0066

In this table the shorthand notation $m\,\mathrm{d}\,n$ has been used where $m\,\mathrm{d}\,n \equiv m \times 10^n$.

Volumes V of spheres and oblate spheroids are computed from one or other of the following expressions:

$$V = \left\| \begin{array}{lll} \frac{4}{3}\pi a^3 & \text{if} & c=a \\ \frac{4}{3}\pi a^2 c & \text{if} & a > c \end{array} \right. \tag{6.30}$$

Table 6.6 was compiled by substituting data on the radii (listed in km and Earth radius units) and oblateness from the JPL D405 ephemeris and *The Astronomical Almanac Online*[39] into Equations (6.29) and (6.30). The planets Mercury and Venus, and the dwarf planet Pluto, were treated as spheres, whilst Earth to Neptune inclusive were treated as oblate spheroids. In the table, column 2 lists the planetary radii in both m (SI unit) and km; column 3 gives the radii in Earth radius units; columns 4 and 5 the planetary surface areas in m^2 and Earth surface area units, and columns 6 and 7 planetary volumes in m^3 and relative to the volume of the Earth.

[39] http://asa.usno.mil/index.html, page K7, 2010.

6.7 Summary and recommendations

6.7.1 Summary

The SI unit of length, the metre, was originally determined over 200 years ago to be one ten millionth part of a meridian arc on the Earth's surface running from the equator to the pole. This length was represented by a pure platinum rod housed under controlled conditions. The precise length of the metre is now fixed by the definition of the speed of light.

The metre and its prefixed derivative, the kilometre, are normally used for measurements of length on the Earth and in near-Earth space (e.g., the distance to the Moon). The radius of the Earth is often used as a longer comparative unit in studies of planetary bodies in the Solar System.

The astronomical unit is used within the Solar System for the computation of planetary ephemerides. Originally defined in terms of the Earth–Sun distance, a revised definition is now the subject of vigorous debate. Whilst adequate as a distance measure in the Solar System, the astronomical unit is too small for interstellar distances, where the parsec and light year are commonly used.

The linear motion of celestial bodies is measured in either the SI derived unit, the $m \cdot s^{-1}$ or, more usually, the $km \cdot s^{-1}$. Other astronomical velocity-related units include red shifts and look-back times.

Acceleration is measured in SI units of $m \cdot s^{-2}$ or the older cgs unit of $cm \cdot s^{-2}$. Examples of the use of this unit as a measure of the surface gravity of bodies in the Solar System is given.

Areas and volumes have SI units m^2 and m^3, respectively. They are used for all celestial bodies, though km^2 and km^3 are common, as is the use of the Earth's radius, surface area and volume as a larger unit for comparative measurements. For larger objects, the Sun's radius, surface area and volume are used.

6.7.2 Recommendations

Wherever possible the SI units, m, $m \cdot s^{-1}$, $m \cdot s^{-2}$, m^2 and m^3 should be used. Until more accurate distance measurements in metres are available, particularly for objects in the outer Solar System, it is appropriate and sensible to go on using astronomical units that depend solely on measures of angle and time. The IAU recommends the use of the parsec, though the light year may prove easier to define, given that it does not rely on the astronomical unit but only on the defined speed of electromagnetic radiation and a defined period of time (one Julian year).

SI units of velocity are routinely used in either the $m \cdot s^{-1}$ form or, for greater velocities, the $km \cdot s^{-1}$ form, and this should continue. For cosmological purposes, red shifts and look-back times can be of more use than simple distance measures.

7
Unit of mass (kilogram)

7.1 SI definition of the kilogram

The **kilogram** is the unit of mass; it is equal to the mass of the international prototype of the kilogram.

The dimension of mass is $[M]$, its unit is the **kilogram** and its symbol is **kg**.

7.1.1 The International Prototype Kilogram

The original definition (1795) of the kilogram was a mass equal to that of a cubic decimeter of pure air-free water at the temperature of melting ice (273.15 K). This was altered four years later to the mass of water in the same volume but at the temperature at which water has its maximum density (which occurs at 277.13 K; Kaye & Laby, 1959). An all-platinum prototype with the same mass as the cubic decimeter of water was manufactured the same year and designated the *Kilogramme des Archives*. The current standard kilogram mass, a cylindrical platinum–iridium alloy, made in 1879 and accepted as the standard since 1889, is known as the International Prototype Kilogram, (IPK). It is now the only SI standard which is a manufactured artifact.[40] The IPK and six replicas are stored at BIPM in a controlled environment. Further copies, known as replicas, were manufactured for distribution to other national metrology laboratories throughout the world.

7.1.2 The stability of the International Prototype Kilogram

As the IPK has, by definition, a mass of one kilogram, it has a zero measurement error. However, when the mass of the IPK is compared with the masses of the replicas, the IPK is apparently losing mass relative to all of them. Conversely, it may be that the replicas are gaining mass relative to the IPK. The apparent loss in mass of the IPK relative to the official copies over a period of just over 100

[40] See http://www.sizes.com/units/BIPM.htm and http://www.sizes.com/units/kilogram.htm

years from 1889 (Girard, 1994) has amounted to $30 \pm 100\,\mu$g or approximately 0.003%.[41]

Such a time-dependent variation is quite obviously unacceptable in the base unit for mass and is of great concern to metrologists worldwide. Active consideration is being given to redefining the unit of mass so that it does not depend on a physical prototype.

7.1.3 Possible future definitions of the kilogram

The CIPM recommended in 2005 that the kilogram be redefined in terms of a fundamental natural constant. It is possible that a new definition will be forthcoming at the 2011 meeting of the CIPM. Some possible ways of arriving at a new definition of the kilogram include the following.

Definition using atomic or subatomic particles, e.g., the number of atoms of carbon or silicon or the number of electrons required to equal the mass of, say, the IPK, at a specified date and time.

Definition by force, using the relationship:

$$F = m \times a \tag{7.1}$$

where F is the force in newtons (kg . m . s^{-2}), m the mass in kg and a the acceleration in m . s^{-2}. Applying a force of 1 N to a mass of 1 kg will result in an acceleration of 1 m . s^{-2}.

Defined by Planck's constant, using the relationship between energy E in joules (m^2 . kg . s^{-2}), frequency ν, in hertz (s^{-1}) and Planck's constant h, in J . s:

$$E = h\nu \tag{7.2}$$

The energy equivalent of a mass of 1 kg is given by:

$$E = mc^2 \tag{7.3}$$

Now rearrange Equation (7.2) and substitute the values for h, E, m (= 1 kg) and c (= 299 792 458 m . s^{-1}) to find ν:

$$\begin{aligned} \nu = \frac{E}{h} &= \frac{(299\,792\,458)^2}{6.626\,068\,96 \times 10^{-34}} \\ &= 1.356 \times 10^{50}\,\text{Hz} \end{aligned} \tag{7.4}$$

It is not practical to measure such a high frequency directly.

41 www.french-metrology.com/en/feature/watt-balance.asp

In a paper on redefining the kilogram, Mills *et al.* (2005) have proposed six separate revised definitions, three that fix the value of the Planck constant and three that fix the value of the Avogadro constant. An example of each type follows.

The kilogram is the mass of a body at rest whose equivalent energy corresponds to frequency of exactly $[(299\,792\,458)^2 / (6\,626\,069\,311)] \times 10^{43}$ Hz.

The kilogram is the mass of a body at rest such that the value of the Avogadro constant is exactly $6.022\,141\,527 \times 10^{23}\,\text{mol}^{-1}$.

Until a final decision is made about the revised formal definition by the CIPM the definition of the kilogram remains as set out at the beginning of this chapter.

7.2 The constant of gravitation

The gravitational force F, with which two bodies of masses M and m attract one another is proportional to the product of their individual masses and the inverse square of their distance r apart. This relationship may be expressed mathematically as:

$$F = G\frac{Mm}{r^2} \tag{7.5}$$

where G, the constant of proportionality, is known as the gravitational constant. The dimension of G is $\mathbf{[L]^3\,.\,[M]^{-1}\,.\,[T]^{-2}}$ and its SI unit is $\mathbf{m^3\,.\,kg^{-1}\,.\,s^{-2}}$. The gravitational constant is very difficult to measure and the accuracy of its measurement has not improved greatly since it was first determined in the late eighteenth century by Henry Cavendish. A recent determination by Fixler *et al.* (2007) using a gravity gradiometer based on atom interferometry yielded a value of $6.693 \times 10^{-11}\,\text{m}^3\,.\,\text{kg}^{-1}\,.\,\text{s}^{-2}$, with both the standard error about the mean value of $\pm 0.027 \times 10^{-11}\,\text{m}^3\,.\,\text{kg}^{-1}\,.\,\text{s}^{-2}$ and a systematic error of $\pm 0.021 \times 10^{-11}\,\text{m}^3\,.\,\text{kg}^{-1}\,.\,\text{s}^{-2}$ given. The CODATA 2006 value is slightly different at $G = 6.674\,28 \times 10^{-11}\,\text{m}^3\,.\,\text{kg}^{-1}\,.\,\text{s}^{-2}$.

The gravitational constant is related to the Gaussian gravitational constant k by:

$$G = k^2 \tag{7.6}$$

The dimension of k is:

$$\dim(k) = [L]^{\frac{3}{2}}\,.\,[M]^{-\frac{1}{2}}\,.\,[T]^{-1} \tag{7.7}$$

and its value is:

$$k = 0.017\,202\,098\,95\,(A)^{\frac{3}{2}}\,.\,(M_{\odot})^{\frac{-1}{2}}\,.\,(d)^{-1} \tag{7.8}$$

when A is distance in astronomical units, $M_{\odot}$ is mass in solar mass units and d is time measured in days of length 86 400 SI seconds.

7.2.1 Values of G measured in some other systems of units

The dimension of G is:

$$\dim(G) = [L]^3 . [M]^{-1} . [T]^{-2} \tag{7.9}$$

and its value in SI units is $6.673 \times 10^{-11}\, \text{m}^3 . \text{kg}^{-1} . \text{s}^{-2}$. To convert G to other systems of units it is only necessary to determine the appropriate constant of proportionality between the systems.

$$G(\text{other units}) = [k_L . L]^3 . [k_M . M]^{-1} . [k_T . T]^{-2} \times G_{\text{SI}} \tag{7.10}$$

Example 1: determine the value of G in cgs units
$k_L = 100$; $k_M = 1000$; $k_T = 1$, where k_L is the number of centimetres in a metre, k_M is the number of grams in a kilogram and $k_T = 1$, since the unit of time, the second, is the same in both systems.

$$\begin{aligned} G_{\text{cgs}} &= [100]^3 . [1000]^{-1} \times G_{\text{SI}} \\ &= 1000 . G_{\text{SI}} \\ &= 6.693 \times 10^{-8}\, \text{cm}^3 . \text{g}^{-1} . \text{s}^{-2} \end{aligned} \tag{7.11}$$

Example 2: determine the value of G in IAU astronomical units
Firstly, compute the number of SI units in each corresponding type of IAU unit. $k_L = 1/(1.495\,978\,71 \times 10^{11})$; $k_M = 1/(1.989 \times 10^{30})$; $k_T = 1/86\,400$ where k_L is the number of astronomical units in a metre, k_M is the number of solar masses in a kilogram and k_T is the number of julian days in an SI second, thus:

$$\begin{aligned} G_{\text{IAU}} &= \left[\frac{1}{1.495\,978\,71 \times 10^{11}}\right]^3 . \left[\frac{1}{1.989 \times 10^{30}}\right]^{-1} . \left[\frac{1}{86\,400}\right]^{-2} . G_{\text{SI}} \\ &\simeq 4.435 \times 10^6 . G_{\text{SI}} \\ &= 2.968 \times 10^{-4} (A)^3 . (M_\odot)^{-1} . (d)^{-2} \end{aligned} \tag{7.12}$$

Example 3: G using the parsec as a unit of length
G_{gal}, which could be of use for dynamical studies in the Milky Way Galaxy, uses the parsec as the unit of length, the solar mass as the unit of mass and the $\text{km} . \text{s}^{-1}$ as the unit of velocity. In this case the dimensional equation is somewhat different:

$$\dim(G_{\text{gal}}) = [k_P . L] . [k_M . M]^{-1} . [k_V . L]^2 . [T]^{-2} \tag{7.13}$$

where $k_P = 1/(3.085\,677\,6 \times 10^{16})$; $k_M = 1/(1.989\,10^{30})$; $k_V = 1/1000$, with k_P being the number of parsecs in a metre, k_M the number of solar masses in a kilogram, k_V the number of kilometres in a metre and the second being common to both systems, then

$$\begin{aligned} G_{\text{gal}} &= \left[\frac{1}{3.085\,677\,6 \times 10^{16}}\right] \cdot \left[\frac{1}{1.989 \times 10^{30}}\right]^{-1} \cdot \left[\frac{1}{1000}\right]^{2} . G_{\text{SI}} \\ &\simeq 6.446 \times 10^{7} . G_{\text{SI}} \\ &= 4.314 \times 10^{-3}\,(\text{pc}) . (M_{\odot})^{-1} . (\text{km} . \text{s}^{-1})^{2} \end{aligned} \tag{7.14}$$

7.2.2 *Standard gravitational parameter*

The standard gravitational parameter μ, is the product of the Newtonian gravitational constant G, and the mass of a given celestial body. If that body is the Sun then the product $G . M_{\odot}$, is known as the heliocentric gravitational constant, or if the body is the Earth, then the product $G . M_{\oplus}$ is known as the geocentric gravitational constant.

From *The Astronomical Almanac Online*[42] the values for the heliocentric and geocentric gravitational constants are given, in SI units, as:

$$\mu_{\odot} = G . M_{\odot} = 1.327\,124\,420\,76 \times 10^{20}\,\text{m}^{3} . \text{s}^{-2} \tag{7.15}$$

$$\mu_{\oplus} = G . M_{\oplus} = 3.986\,004\,418 \times 10^{14}\,\text{m}^{3} . \text{s}^{-2} \tag{7.16}$$

Note that the accuracy with which the values of μ may be expressed far exceeds that of the constant of gravitation as determined in the laboratory. The best current values for G have standard deviations about the mean measured value of approximately $\pm 0.01\,\%$ or 1 part in 10^{4}, whereas the heliocentric gravitational constant is known to approximately 1 part in 10^{10}. Hence it is customary when computing the values for semimajor axes of Solar System bodies to use IAU units rather than SI units via the relationship (Kepler's third law):

$$a = \sqrt[3]{\left(\frac{G . M_{\odot} . P^{2}}{4 . \pi^{2}}\right)} = \sqrt[3]{\left(\frac{k^{2} . M_{\odot} . P^{2}}{4 . \pi^{2}}\right)} = \sqrt[3]{\left(\frac{\mu . P^{2}}{4 . \pi^{2}}\right)} \tag{7.17}$$

where a is the semimajor axis of the orbit, P is the orbital period, $M_{\odot}$ is the mass of the Sun, G is the Newtonian gravitational constant, k is the Gaussian gravitational constant and μ is the standard gravitational parameter. Planetary ephemerides are currently computed using k with distances in astronomical units and masses in inverse mass units $M_{\odot} / M_{P}$.

42 http://asa.usno.mil/index.html, page K7, 2010.

The standard gravitational parameter may be transformed from SI units to IAU units in the following way. The dimension of μ is:

$$\dim(\mu) = [L]^3 . [T]^{-2} \tag{7.18}$$

so

$$\mu_{\mathrm{IAU}} = [k_L . L]^3 . [k_T . T]^{-2} . \mu_{\mathrm{SI}} \tag{7.19}$$

where k_L is the number of astronomical units in a metre and k_T is the number of julian days in an SI second, hence:

$$\begin{aligned} \mu_{\mathrm{IAU}} &= \left[\frac{1}{1.495\,978\,71 \times 10^{11}}\right]^3 . \left[\frac{1}{86400}\right]^{-2} . \mu_{\mathrm{SI}} \\ &\simeq 2.230 \times 10^{-24} . \mu_{\mathrm{SI}} \\ &= 2.959 \times 10^{-4}\,(\mathrm{A})^3 . (\mathrm{d})^{-2} \end{aligned} \tag{7.20}$$

7.3 Masses of astronomical bodies

The masses of bodies in the Solar System may be determined using Kepler's third law and Newton's second law, augmented by direct measurements for the Earth and helioseismology for the Sun. For binary stars, Kepler's third law is again invoked. For single stars, the empirically determined relationship between mass and luminosity may be used, as well as asteroseismology, and for stars which act as the lensing star in a microlensing event, general relativity allows an estimate of masses through the observed alignment geometry of the two stars (source and lens). Galaxies present more of a problem since the use of Kepler's third law is inadequate when the possibility of dark matter and even dark energy have to be taken into account.

7.3.1 The solar mass

By rearranging Equation (7.17), the mass of the Sun, $M_\odot$, may be expressed approximately in SI units as:

$$M_\odot \simeq \frac{4 . \pi^2 . a^3}{G . P^2} \tag{7.21}$$

where a is the semimajor axis of a planetary orbit in metres, P is the period of revolution in SI seconds, and G is the Newtonian gravitational constant in $\mathrm{m}^3 . \mathrm{kg}^{-1} . \mathrm{s}^{-2}$. Substituting the values for a and P that approximate for the Earth and making the assumption that the mass of the Earth is small in comparison with that of the Sun gives:

$$M_\odot \simeq \frac{4.\pi^2.(1.495\,978\,714\,64 \times 10^{11})^3}{6.673 \times 10^{-11}.(365.25 \times 86\,400)^2}$$
$$\simeq 1.989 \times 10^{30}\,\text{kg} \qquad (7.22)$$

Note that if IAU units are used, then $P = 365.25$ d, $a = 1$ (A), and $G = 2.968 \times 10^{-4}\,(\text{A})^3.(M_\odot)^{-1}.(\text{d})^{-2}$, so that:

$$M_\odot \simeq \frac{4.\pi^2.1^3}{2.968 \times 10^{-4}.365.25^2}$$
$$\simeq 0.997$$
$$\simeq 1 \qquad (7.23)$$

The mass of the Sun may also be determined from observations of the orbital velocity and position of a planet at a particular time. If the measured velocity of the planet is v_P, and its radius vector is R_P, then the approximate mass of the Sun may be calculated from:

$$M_\odot = \frac{v_P^2.R_P}{G} \qquad (7.24)$$

where G is the Newtonian constant of gravitation.

Example: determine the approximate mass of the Sun from observations of the planet Venus

The heliocentric coordinates (x, y, z) in astronomical units and velocity components $(\dot{x}, \dot{y}, \dot{z})$ in astronomical units per day of Venus at JD = 2 449 720.5 are given in the *The Astronomical Almanac* (1995) as:

$$x = -0.541\,0794 \qquad y = +0.417\,3425 \qquad z = +0.221\,9973$$
$$\dot{x} = -0.013\,358\,66 \qquad \dot{y} = -0.014\,313\,22 \qquad \dot{z} = -0.005\,593\,58$$

The length of the radius vector Sun–Venus, R, is given by:

$$R = \sqrt{x^2 + y^2 + z^2} \qquad (7.25)$$

and the rate of change of the radius vector, v, by:

$$v = \sqrt{\dot{x}^2 + \dot{y}^2 + \dot{z}^2} \qquad (7.26)$$

Substituting the coordinates and velocity components from above in Equations (7.25) and (7.26) gives values for the radius vector and its time derivative of $R = 0.718\,487\,63$ (A) and $v = 0.020\,361\,98\,(\text{A}).(\text{d})^{-1}$. Using the value of G in

IAU units plus the values for $R_P = R$ and $v_P = v$ in Equation (7.24) yields a value for the mass of the Sun of $M_\odot \simeq 1.004$ (in solar mass units).

To determine the mass of the Sun in SI units multiply R by $1.495\,978\,714\,64 \times 10^{11}$ (the number of metres in an astronomical unit) and divide by the number of SI seconds in a julian day (86 400). Insert the adjusted numbers into Equation (7.24) to give $M_\odot \simeq 2.002 \times 10^{30}$ kg.

The solar mass in general relativity

In general relativity, the mass of the Sun is sometimes given in the form of a unit of length $(M_\odot)_L$ or time $(M_\odot)_T$ as follows:

$$(M_\odot)_L = \frac{G\,.\,M_\odot}{c^2} = \frac{1.327\,124\,420\,76 \times 10^{20}}{(2.997\,924\,58 \times 10^8)^2} = 1.476\,625\,061 \times 10^3\,\text{m} \tag{7.27}$$

$$(M_\odot)_T = \frac{G\,.\,M_\odot}{c^3} = \frac{1.327\,124\,420\,76 \times 10^{20}}{(2.997\,924\,58 \times 10^8)^3} = 4.925\,491\,025 \times 10^{-6}\,\text{s} \tag{7.28}$$

Note that the solar mass units are now quoted in metres and seconds, which may be verified using dimensional analysis.

$$\dim((M_\odot)_L) = \frac{\dim(G)\,.\,\dim(M_\odot)}{\dim(c^2)} = \frac{[L]^3\,.\,[M]^{-1}\,.\,[T]^{-2}\,.\,[M]}{[L]^2\,.\,[T]^{-2}} = [L] \tag{7.29}$$

$$\dim((M_\odot)_T) = \frac{\dim(G)\,.\,\dim(M_\odot)}{\dim(c^3)} = \frac{[L]^3\,.\,[M]^{-1}\,.\,[T]^{-2}\,.\,[M]}{[L]^3\,.\,[T]^{-3}} = [T] \tag{7.30}$$

7.3.2 Mass of the Earth

The earliest attempts at determining the mass of the Earth relied on regularly shaped mountains. A plumb bob is gravitationally attracted by the mountain so that the plumb line no longer points directly to the zenith. Measurements of this small deviation and the estimated mass of the mountain leads to an approximate value of the mass of the Earth. Laboratory-based determinations later used torsion balances and gravity balances (Spencer-Jones, 1956).

The advent of artificial satellites in orbit about the Earth, which obey Kepler's third law, allow the same dynamical method to be employed in the determination of the Earth's mass as is used for the Sun. If, in Equation (7.21), the mass of the Earth $M_{\oplus}$, is substituted for the mass of the Sun, and the semimajor axis a_S and the orbital period P_S of the artificial satellite for those of the Earth, then:

$$M_{\oplus} = \frac{4.\pi^2.a_S^3}{G.P_S^2} \tag{7.31}$$

Example: The International Space Station and the mass of the Earth

By way of an example it is possible to determine approximately the mass of the Earth using observations of the International Space Station. Measurements of the orbital period in minutes and the height of the apogee point of the orbit above the surface of the Earth in kilometres are to be found on the satellite database of astrosat.net.[43] The semimajor axis length of the satellite's orbit is taken to be the sum of the Earth's equatorial radius and the satellite's apogee distance and the listed orbital period multiplied by 60 to convert to seconds of time. The value G_{SI} is substituted in Equation (7.31) for G to give:

$$\begin{aligned} M_{\oplus} &= \frac{4.\pi^2.(6.732\times10^6)^3}{6.673\times10^{-11}.(5.491\times10^3)^2} \\ &= 5.97\times10^{24}\,\text{kg} \end{aligned} \tag{7.32}$$

which, allowing for the very basic analysis performed, is a reasonable approximation to the mass of the Earth given in *The Astronomical Almanac Online* as $5.972\,198\,6\times10^{24}$ kg.[44]

For a more detailed account of how to determine the mass of the Earth or any other planet taking into account the oblateness of the planet see, e.g., Blanco & McCuskey (1961).

7.3.3 Masses of Solar System objects

The masses of all the major planets and some of the dwarf planets in the Solar System may be determined using observations of their oblateness in conjunction with a detailed study of the orbits of one or more of their satellites (either natural or artificial). Other than using the SI kilogram, the masses of the Sun and the planets Earth and Jupiter are also commonly used as units in planetary and exoplanetary studies.

[43] See www.satview.com.br/us/track_lista_sat.php
[44] http://asa.usno.mil/index.html, page K7, 2010.

Table 7.1. *Masses, in kilograms, solar mass units, reciprocal solar mass units, Jovian mass units and Earth mass units of the Sun, planets, dwarf planets, satellites and a comet in the Solar System*

Name	Mass				
Star	kg	$M_\odot$	$1/M_\odot$	M_J	$M_\oplus$
Sun	1.9884 d 30	1	1	1.0473 d 3	3.3294 d 5
Planet					
Mercury	3.3010 d 23	1.6601 d − 7	6.0236 d 6	1.7387 d − 4	0.0553
Venus	4.8673 d 24	2.4478 d − 6	4.0852 d 5	2.5637 d − 3	0.8150
Earth	5.9721 d 24	3.0035 d − 6	3.3295 d 5	3.1457 d − 3	1
Mars	6.4169 d 23	3.2272 d − 7	3.0987 d 6	3.3800 d − 4	0.1074
Jupiter	1.8985 d 27	9.5479 d − 4	1.0473 d 3	1	317.891
Saturn	5.6846 d 26	2.8589 d − 4	3.4979 d 3	2.9942 d − 1	95.184
Uranus	8.6818 d 25	4.3662 d − 5	2.2903 d 4	4.5730 d − 2	14.5371
Neptune	1.0243 d 26	5.1514 d − 5	1.9412 d 4	5.3953 d − 2	17.1512
Dwarf Planet					
Ceres	9.3455 d 20	4.7 d − 10	2.1 d 9	4.9 d − 7	1.6 d − 4
Pluto	1.4707 d 22	7.3964 d − 9	1.3520 d 8	7.7467 d − 6	2.4626 d − 3
Eris	1.9884 d 20	1 d − 10	1 d 10	1.0 d − 7	3.3294 d − 5
Satellite					
Moon	7.3458 d 22	3.6943 d − 8	2.7068 d 7	3.8693 d − 5	1.230 d − 2
Ganymede	1.4818 d 23	7.4522 d − 8	1.3419 d 7	7.805 d − 5	2.48 d − 2
Europa	4.7994 d 22	2.4137 d − 8	4.1430 d 7	2.528 d − 5	8.04 d − 3
Comet					
Halley	5 d 14	2.4 d − 16	4 d 15	2.6 d − 13	8.4 d − 11

In this table the shorthand notation $m\,\mathrm{d}\,n$ has been used where $m\,\mathrm{d}\,n \equiv m \times 10^n$.

Examples of the masses of the planets, dwarf planets, satellites and Halley's comet are given in Table 7.1. The values were computed from data given in Cox (2000) (for the Moon), Brown & Schaller (2007) (for Eris) and *The Astronomical Almanac Online*.

It should be remembered that until the value of the solar mass in kilograms is greatly improved, the relative masses of the planets in solar mass units are much more accurate. In Table 7.1, the masses in kilograms of the various objects were derived by multiplying the mass of the body in solar mass units by one solar mass in kilograms which, being given to five significant figures, immediately restricts all the other derived values to being no better than five significant figures, and generally rather worse. The values given for the reciprocal solar masses of these same objects range from seven significant figures for Uranus and Neptune up to 12 significant figures for the Earth, which explains why IAU units are to be preferred for the computations of ephemerides at present.

7.3.4 Stellar masses

The masses of double stars that are gravitationally bound may have their component masses determined using Kepler's third law and Newton's laws of motion.

In a binary star system each component describes an elliptical orbit about the common centre of gravity (the barycentre), as is illustrated in Figure 7.1. If the binary star is relatively close to the Sun then each component star may be individually observed and plotted on the plane of the sky to construct an orbit. If the binary is much further away the component stars may not be resolved as individual stars but may be detected via their composite spectrum, in which the spectral lines generated by each star will move relative to each other. This type of binary is known as a double-line spectroscopic binary, a special case of which occurs when one member star is so much fainter than the other that its spectral lines are invisible but its presence is still detectable by the effect it has on the spectral lines of the brighter component, which shift in a regular fashion. Such a binary is known as a single-line spectroscopic binary. For spectroscopic binaries of either type whose orbital plane lies along the line of sight to the binary, regular eclipses occur as one component star passes in front of the other. These objects are known as eclipsing binaries.

For a binary star, the sum of the masses of the components may be derived from Kepler's third law:

$$M_{\mathrm{A}} + M_{\mathrm{B}} = \frac{4.\pi^2.a^3}{G.P^2} \tag{7.33}$$

where a is the semimajor axis of the absolute elliptical orbit in metres, P is the orbital period of the system in SI seconds and M_{A} and M_{B} are the masses of the individual stars in kilograms, with G the constant of gravitation in SI units. So, if the orbit of the binary is well defined, it is a simple matter to substitute appropriate

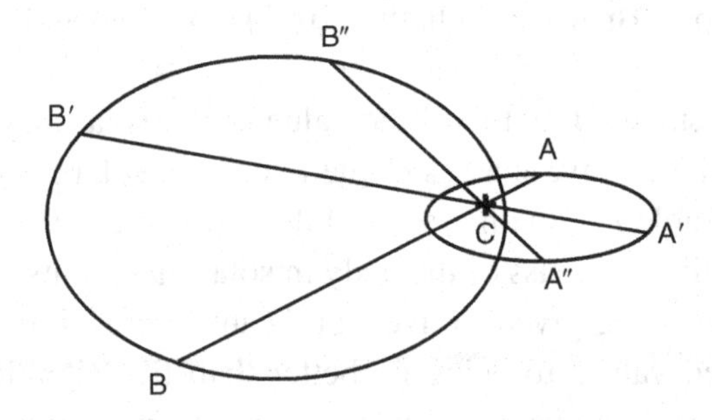

Figure 7.1. The orbits of a pair of stars, A and B, in a binary system about their centre of mass, C, the barycentre. Star A moves from A through A′ to A″ whilst star B moves from B through B′ to B″. The barycentric distances BC and CA are equal to r_{B} and r_{A}, respectively.

values into Equation (7.33) to obtain the total mass of the binary system. To obtain the individual masses, consider the binary star orbits shown in Figure 7.1. The distance from star A to the barycentre C, is r_A, and likewise star B is distance r_B from the barycentre. Distance $\overrightarrow{AB} = r$, so $r = r_A + r_B$. The total mass of the system is M, where $M = M_A + M_B$ and $M_A . r_A = M_B . r_B$. Using Newton's second law of motion balancing gravitational and centripetal force gives:[45]

$$\frac{G . M_A . M_B}{r^2} = \frac{M_A . \dot{r}_A^2}{r_A} \tag{7.34}$$

where $\dot{r}_A$ is the orbital speed of star A, which may be measured spectroscopically or simply taken as the mean value defined by:

$$\dot{r}_A = \frac{2 . \pi . r_A}{P} \tag{7.35}$$

Combining the various expressions above produces the relationship:

$$M_A = \frac{M . (r - r_A)}{r} \tag{7.36}$$

The value for the mass of star B is derived from:

$$M_B = M - M_A \tag{7.37}$$

If IAU units are used with mass in solar mass units, distance in astronomical units and time in days of 86 400 SI seconds, Equation (7.33) may be rewritten as:

$$M_{IAU} = \frac{4 . \pi^2 . a_{IAU}^3}{G_{IAU} . P_{IAU}^2} \tag{7.38}$$

which gives the mass of the binary system in solar mass units. Equations (7.34) to (7.37) may similarly be reworked.

Example: determine the total mass of the Sirius binary system in both IAU and SI units
(1) IAU units

Parallax: 379 mas
Distance: $\frac{206\,264.806\,25}{0.379} = 5.442\,343 \times 10^5$ au
Orbital semimajor axis (in radians): $\frac{7.5}{206\,264.806\,25}$ rad
Orbital semimajor axis (in astronomical units): $\frac{7.5 \times 5.442\,343 \times 10^5}{206\,264.806\,25} = 19.789$ au

[45] For more detail and a worked example, see http://outreach.atnf.csiro.au/education/senior/astrophysics/binary_mass.html

Constant of gravitation: $2.968 \times 10^{-4}\,\mathrm{A}^3\,.M_{\odot}^{-1}\,.\mathrm{d}^{-2}$

Total mass of the Sirius system: $\frac{4\,.\,\pi^2 \times (19.789)^3}{2.968 \times 10^{-4} \times (1.829\,537 \times 10^4)^2} = 3.08\,M_{\odot}$

(2) SI units

Distance: $8.141\,63 \times 10^{16}$ m
Period: $1.580\,72 \times 10^9$ s
Orbital semimajor axis: $2.960\,38 \times 10^{12}$ m
Constant of gravitation: $6.673 \times 10^{11}\,\mathrm{m}^3\,.\,\mathrm{kg}^{-1}\,.\,\mathrm{s}^{-2}$
Total mass of the Sirius system: 6.143×10^{30} kg ($\simeq 3.09\,M_{\odot}$)

The Sirius binary system has the catalogue number WDS06451-1643 in the online *Washington Double Star Catalog*. The orbital and distance data were obtained from this catalogue.[46]

It is not uncommon in binary-star research to measure the orbital period of the system in years rather than days, which alters the approximate value of $4\,.\,\pi^2\,/\,G$ to unity, as follows:

$$P_{\mathrm{d}} = 365.25 \times P_{\mathrm{y}}$$

$$M = \frac{4\,.\,\pi^2\,.\,a_{\mathrm{IAU}}^3}{G_{\mathrm{IAU}}\,.\,(P_{\mathrm{y}} \times 365.25)^2}$$

$$= 0.997 \times \frac{a_{\mathrm{IAU}}^3}{P_{\mathrm{y}}^2}$$

$$\simeq \frac{a_{\mathrm{IAU}}^3}{P_{\mathrm{y}}^2} \qquad (7.39)$$

So when the semi-major axis of the binary-star orbit is measured in astronomical units, its period is measured in years and the value of $4\,.\,\pi^2\,/\,G$ is set equal to 1, then the total mass of the binary system is measured in solar mass units.

Asteroseismology

Asteroseismology may be defined as the study of the internal structure of individual stars through the analysis of their pulsation periods and the interpretation of the derived frequency spectra. Stellar masses and radii can be determined to a much higher precision than by any other method.[47]

This branch of astronomy has matured in recent times, with the development of high-speed photometers and the establishment of networks of observational sites throughout the world (e.g., the WET (Whole Earth Telescope) collaboration). The

[46] See http://ad.usno.navy.mil/wds/
[47] See http://kepler.nasa.gov/Science/about/RelatedScience/Asteroseismology/

analysis of the power spectrum of the light curve (Winget *et al.*, 1994) allows direct estimations to be made of the masses of individual stars. For the pulsating DB white dwarf GD358, Bradley (1993) derived a mass of $0.61 \pm 0.03\ M_\odot$.

Asteroseismological reduction techniques have also been applied to observations made of relatively bright main-sequence and subgiant branch stars using spectrographs. Bedding *et al.* (2006) used such observations to determine the mass of the metal-poor subgiant star ν Indi to be $0.85 \pm 0.04\ M_\odot$. Ferdman *et al.* (2010) used radio telescopes in the USA, Australia and France to make extensive observations of the intermediate-mass binary pulsar PSR J1802-2124, which yielded masses for the pulsar component of $1.24 \pm 0.11\ M_\odot$ and for the white-dwarf component of $0.78 \pm 0.04\ M_\odot$. Such measurements and analysis provide important independent estimates of stellar masses using a method other than those of Kepler/Newton dynamics.

Some examples of stellar masses determined from dynamical and asteroseismological techniques are given in Table 7.2, which was compiled, in part, from material to be found in Cox (2000), Ferdman *et al.* (2010), Winget *et al.* (1994) and Bedding *et al.* (2006). It is worth noting that currently there is no prefix in SI that is large enough for stellar masses, even the smallest mass in Table 7.2 would be given as 3.182×10^8 Yg (3.182×10^8 yottagrams). The additional, unofficial, prefixes proposed by Mayes (1994) would describe the mass as 318.2 Sg (318.2 sansagrams). (Again, remembering that the prefix is applied to the mass unit gram and not to the SI unit kilogram.)

7.3.5 *Mass loss and gain*

The masses of celestial bodies may be either increased or decreased due to various physical processes, such as nuclear reactions converting mass into energy, stellar winds, accretion and collisions.

The dimension of mass change is **[M] . [T]**$^{-1}$ its unit is the **kilogram per second** and its symbol **kg . s**$^{-1}$. Non-SI units include the kg . y^{-1}, $M_\odot$. y^{-1} and the cgs version g . s^{-1}.

Example of mass gain: According to Love & Brownlee (1993), the infall of dust and small meteoroids on to the Earth amounts to some 4×10^7 kg . y^{-1}, equivalent to approximately 1.3 kg . s^{-1}.

Examples of mass loss: in the region near the centre of the Sun, energy is generated by the conversion of hydrogen nuclei into helium nuclei. This reaction results in a loss of mass amounting to 4.3×10^9 kg . s^{-1}, with a further 1.3×10^9 kg . s^{-1} being lost via the solar wind. A total mass loss of 5.6×10^9 kg . s^{-1}, equivalent in IAU units to $2.4 \times 10^{-16}\ M_\odot$. d^{-1}, or $8.8 \times 10^{-14}\ M_\odot$. y^{-1}.

Table 7.2. *Masses of binary-star components determined dynamically and single stars determined using asteroseismological methods*

Star identifier	Spectral type	Mass kg	$M_\odot$
Sirius A	A1V	4.535 d 30	2.28
Procyon A	F5IV–V	3.361 d 30	1.69
PSR J1802-2124 A	Pulsar	2.466 d 30	1.24
α Cen A	G2V	2.148 d 30	1.08
SUN	G2V	1.989 d 30	1.00
Sirius B	DA	1.949 d 30	0.98
70 Oph A	K0V	1.790 d 30	0.90
α Cen B	K0V	1.750 d 30	0.88
ν Ind	G0IV	1.691 d 30	0.85
70 Oph B	K4V	1.293 d 30	0.65
GD358	DBV	1.213 d 30	0.61
Procyon B	WD	1.193 d 30	0.60
Krüger 60 A	dM4	5.370 d 29	0.27
Krüger 60 B	dM6	3.182 d 29	0.16

The recently discovered super-Earth exoplanet CoRoT-7b is located very close to its star and is subject to intense EUV radiation, causing a mass loss from the planet of approximately 10^{11} g . s^{-1}, or 10^{8} kg . s^{-1} in SI units, according to Valencia *et al.*[48]

7.4 Density

Density may be defined as the mass or quantity of matter of a substance per unit volume.

The dimension of density is $[M].[L]^{-3}$, its derived unit is the **kilogram per cubic metre** and its symbol, **kg . m^{-3}**.

Density unit conversion

Units of density other than the kg . m^{-3} in common use in astronomy include the gram per cubic centimetre, g . cm^{-3}, and the solar mass per cubic parsec, $M_\odot$. pc^{-3}. Conversion factors between the various units are given below as worked examples.

[48] Reported by D. Valencia, M. Ikoma, T. Guillot and N. Nettlemann at the 41st Lunar and Planetary Science Conference (2010) in a paper entitled 'Composition and fate of short-period super-Earth's: The case of CoRoT-7b.'

Table 7.3. *Densities of various celestial objects*

Celestial object	Mean density $\mathrm{kg\,.\,m^{-3}}$	$\mathrm{g\,.\,cm^{-3}}$	$M_{\odot}\,.\,\mathrm{pc}^{-3}$	Notes
Sun	1409	1.409	–	mean density
Earth	5514.8	5.515	–	
Jupiter	1330	1.33	–	
Saturn	700	0.70	–	
Moon	3341	3.341	–	mean density
Asteroids	2250	2.25	–	mean of estimated range of densities
Halley's comet	650	0.65	–	mean of estimated range of densities
Interstellar matter	2.71 d – 21	2.71 d – 24	0.04	
Main sequence stars	3.38 d – 21	3.38 d – 24	0.050	mass density in the solar neighbourhood

In this table the shorthand notation $m\,\mathrm{d}n$ has been used where $m\,\mathrm{d}n \equiv m \times 10^{n}$.

Example 1: cgs unit conversions

$$1\,\mathrm{g\,.\,cm^{-3}} = \left(\frac{1}{1000}\right).\mathrm{kg}\,.\left(\frac{1}{100}\right)^{-3}.\,\mathrm{m^{-3}}$$
$$= 1000\,\mathrm{kg\,.\,m^{-3}} \qquad (7.40)$$

the inverse of which is:

$$1\,\mathrm{kg\,.\,m^{-3}} = 0.001\,\mathrm{g\,.\,cm^{-3}} \qquad (7.41)$$

Example 2: solar mass per cubic parsec unit conversions

$$1M_{\odot}\,.\,\mathrm{pc^{-3}} = (1.989 \times 10^{30})\,.\,\mathrm{kg}\,.(3.085\,667\,6 \times 10^{16})^{-3}\,.\,\mathrm{m^{-3}}$$
$$= 6.769\,980 \times 10^{-20}\,\mathrm{kg\,.\,m^{-3}} \qquad (7.42)$$

the inverse of which is:

$$1\,\mathrm{kg\,.\,m^{-3}} = 1.477\,109 \times 10^{19}\,M_{\odot}\,.\,\mathrm{pc^{-3}} \qquad (7.43)$$

Some examples of the densities of astronomical bodies are given in Table 7.3, with data from Cox (2000) recalculated where necessary.

A form of density unit that is often used is the **column mass density**, whose unit is the $\mathbf{M_{\odot}\,.\,pc^{-2}}$. Column mass density is normally accompanied by a range within which the quoted value is valid. For example, the observed column mass density to

$|z| = 1.1$ kpc of the interstellar medium is approximately 13 $M_{\odot}\,.\,\mathrm{pc}^{-2}$. In SI units, this column mass density would be given as:

$$|z| = 1.1 \times 1000 \times 3.0856676 \times 10^{16}$$
$$= 3.394234 \times 10^{19}\,\mathrm{m} \qquad (7.44)$$
$$\rho = \frac{13 \times 1.989 \times 10^{30}}{(3.0856676 \times 10^{16})^2}$$
$$= 0.027157\,\mathrm{kg}\,.\,\mathrm{m}^{-2} \qquad (7.45)$$

so in SI units the column mass density of the local interstellar medium would be $0.027\,\mathrm{kg}\,.\,\mathrm{m}^{-2}$ within $|z| = 3.394 \times 10^{19}$ m.

7.5 Force

Funk *et al.* (1946) define force as: 'Any cause that produces, stops, changes or tends to produce, stop or change the motion of a body.'

The dimension of force is **[*M*] . [*L*] . [*T*]**$^{-2}$, its derived SI unit is the **newton** with symbol, **N**, or, $\mathbf{kg\,.\,m\,.\,s^{-2}}$.

Another unit still used by astronomers is the dyne, a cgs unit equivalent to $1\,\mathrm{g}\,.\,\mathrm{cm}\,.\,\mathrm{s}^{-2}$. The conversion coefficient between the two systems is derived as follows:

$$1\,\mathrm{kg}\,.\,\mathrm{m}\,.\,\mathrm{s}^{-2} = 1000\,.\,\mathrm{g}\,.\,100\,.\,\mathrm{cm}\,.\,\mathrm{s}^{-2}$$
$$= 10^5\,.\,\mathrm{g}\,.\,\mathrm{cm}\,.\,\mathrm{s}^{-2} \qquad (7.46)$$

or 1 newton $= 10^5$ dynes.

7.5.1 Energy

The energy of a body is its capacity for doing work, where work is defined as the product of the force that moves the body and the distance through which it is moved. The SI units for energy and work are the same.

The dimension of energy is **[*M*] . [*L*]**^{2} **. [*T*]**$^{-2}$, its derived SI unit is the **joule** with symbol, **J**, or **N . m** or $\mathbf{kg\,.\,m^2\,.\,s^{-2}}$.

The cgs unit of energy, the erg, is regularly used by astronomers. Measures in ergs are converted to joules via 1 erg $= 10^{-7}$ J.

The **potential energy** of a system is the energy that the system has due to the relative positions of its component parts.

The **kinetic energy** of a moving body or system is the energy it possesses due to its motion.

The dimensions, units and symbols of both potential and kinetic energy are the same as those for energy.

7.5.2 Power

Power is the rate at which energy is transferred per unit time.

The dimension of power is $[M].[L]^2.[T]^{-3}$, its derived SI unit is the **watt** with symbol **W**, or $\mathbf{J.s^{-1}}$, or $\mathbf{kg.m^2.s^{-3}}$.

The cgs unit of power, the erg . s^{-1}, is still used as, e.g., a measure of luminosity in model stellar atmosphere calculations (see Table G.10 in Irwin, 2007).

7.5.3 Pressure

Pressure is a force applied perpendicularly to a unit area.

The dimension of pressure is $[M].[L]^{-1}.[T]^{-2}$, its derived SI unit is the **pascal** with symbol, **Pa**, or $\mathbf{N.m^{-2}}$, or $\mathbf{kg.m^{-1}.s^{-2}}$.

Astronomers tend to use the cgs unit of pressure, the dyn . cm^{-2}, or the bar, equal to 10^6 dyn . cm^{-2}. The conversion factors from these units to the SI unit of pressure are:

$$1\,\text{bar} = 10^5\,\text{Pa} \qquad 1\,\text{Pa} = 10^{-5}\,\text{bar}$$

$$1\,\text{dyn.cm}^{-2} = 10^{-6}\,\text{bar} \qquad 1\,\text{Pa} = 10\,\text{dyn.cm}^{-2} \tag{7.47}$$

The atmospheric surface pressures on the Sun, some planets, a dwarf planet and a satellite, listed by Cox (2000), are converted to SI units using the appropriate factors from Equations (7.47) and set out in Table 7.4.

The atmospheric pressure given for the Sun refers to a point where the radial optical depth at a wavelength of 500 nm is equal to 1.

7.6 Moments of inertia and angular momentum

The **moment of inertia, I**, of a rigid body about a given axis is the sum of the products $\boldsymbol{m.r^2}$, for all the elements composing the body, where m is the mass of each element and r is its perpendicular distance from the axis concerned.

$$I = \sum m.r^2 \tag{7.48}$$

The dimension of moment of inertia is $[M].[L]^2$, its unit is the **kilogram metre squared** and its symbol is $\mathbf{kg.m^2}$.

Table 7.4. *Surface pressures for a selection of various celestial objects*

Celestial object	Surface atmospheric pressure		
	Pa	$dyn\,.\,cm^{-2}$	bar
Star			
Sun	1.207 d 4	1.207 d 5	1.207 d−1
Planet			
Venus	9 d 6	9 d 7	90
Earth	1 d 5	1 d 6	1
Mars	1 d 3	1 d 4	0.01
Jupiter	3 d 4	3 d 5	0.3
Saturn	4 d 4	4 d 5	0.4
Dwarf planet			
Pluto	8	80	8 d−5
Satellite			
Titan	1.5 d 5	1.5 d 6	1.5

In this table the shorthand notation $m\,\mathrm{d}\,n$ has been used where $m\,\mathrm{d}\,n \equiv m \times 10^n$.

Moments of inertia are often presented in cgs units as $g\,.\,cm^2$ and, for Solar System objects, as $M\,.\,R^2$, where M is the mass of the objects in units of the Earth's mass and R is the object's radius in units of the Earth's radius.

To convert from cgs to SI units: the conversion factor is derived in the following way:

$$I_{\mathrm{SI}} = \left(\frac{1}{1000}\right) . \left(\frac{1}{100}\right)^2 . I_{\mathrm{cgs}}$$
$$= 10^{-7} . I_{\mathrm{cgs}} \tag{7.49}$$

In Cox (2000), the moment of inertia of the Sun is given as $5.7 \times 10^{53}\ g\,.\,cm^2$, which, using the conversion factor in Equation (7.49), becomes $5.7 \times 10^{46}\ kg\,.\,m^2$ in SI units.

To convert from $M_\oplus\,.\,R_\oplus^2$ units: the conversion factor is derived by substituting values for the Earth's mass in kilograms and the Earth's radius in metres:

$$I_{\mathrm{SI}} = (5.972\,198\,6 \times 10^{24}) . (6.378\,136\,6 \times 10^{6})^2 . I_{M_\oplus . R_\oplus^2}$$
$$= 2.429\,527 \times 10^{38} . I_{M_\oplus . R_\oplus^2} \tag{7.50}$$

In Cox (2000), the moment of inertia of the Earth is given as $0.333\,5\ M_\oplus\,.\,R_\oplus^2$ which is readily converted into SI units via Equation (7.50) to equal $8.10 \times 10^{37}\ kg\,.\,m^2$.

The **angular momentum, Iω**, of a celestial body about its rotation axis is defined as the product of its moment of inertia and its angular velocity.

The dimension of angular momentum is $[\boldsymbol{M}].[\boldsymbol{L}]^2.[\boldsymbol{T}]^{-1}$, its unit is the **kilogram metre squared per second**, and its symbol is $\mathbf{kg.m^2.s^{-1}}$.

In Cox (2000), the angular momentum of the Sun is given in cgs units as 1.63×10^{48} g . cm^2 . s^{-1}. The time unit is the same in both cgs and SI units. The conversion factor from cgs to SI units is the same as that given in Equation (7.49), so that in SI units the solar angular momentum is 1.63×10^{41} kg . m^2 . s^{-1}.

7.7 Summary and recommendations

7.7.1 Summary

The SI unit of mass the kilogram is the only remaining base unit that is defined by a prototype, the International Prototype Kilogram (IPK), a block of platinum–iridium of mass 1 kg. Comparison of the mass of the IPK with a set of replicas over more than a century has shown that the IPK is losing mass. Obviously, this is an unacceptable situation for a base unit standard and considerable effort is being expended in trying to arrive at a new definition for the kilogram.

In dynamical astronomy the constant of gravitation is of great importance, being the constant of proportionality relating the magnitude of the gravitational attraction or force between two bodies and their individual masses and separation. Unfortunately, it has proved extremely difficult to measure, which has resulted in a low precision value ($\sim$1 part in 10^4) for the constant. However, the standard gravitational parameter, the product of the constant of gravitation and the solar mass, may be determined with far more precision ($\sim$1 part in 10^{10}).

The masses of celestial objects are determined in a variety of ways using dynamics, asteroseismology, a relationship between mass and luminosity and gravitational microlensing. Astronomical body masses are most commonly given relative to the mass of the Sun, Earth or other Solar System objects.

The SI units used for density, force, energy, power, pressure, moments of inertia and angular momentum are all given and are relevant and simple to use, though are less common than either cgs units, IAU units or versions of IAU units (e.g., Earth mass and radius units rather than solar mass and radius units).

Worked examples of how to transform the various cgs and IAU units into SI units are given throughout the chapter.

7.7.2 Recommendations

The probable change from the IPK to a fundamental particle or energy definition of the SI kilogram should have no effect on the use of the kilogram as the base unit of mass in astronomy.

At present, the lack of precision in the determination of G means that it is sensible to continue using IAU units for planetary dynamics, since only relative masses are available to a high accuracy.

Unconventional units are regularly used to represent mass – including those based on terrestrial and Jovian masses. Cgs units are mainly used with moments of inertia and angular momenta. Atmospheric pressures are regularly quoted in dyn . cm^{-2} or bars instead of pascals, though since the differences between them are only powers of ten it is simple to effect the changeover to SI units. However, should there be a valid reason for not changing, then perhaps an extra entry in the abstract and summary giving the main result of the research programme in SI units should not prove too difficult.

8

Unit of luminous intensity (candela)

8.1 SI definition of the candela

The **candela** is the luminous intensity, in a given direction, of a source that emits monochromatic radiation of frequency 540×10^{12} hertz and that has a radiant intensity in that direction of 1/683 watt per steradian.

The dimension of luminous intensity is [$\boldsymbol{J}$], its unit is the **candela** and its symbol is **cd**.

8.2 Radiometry and photometry

Radiometry is the measurement of electromagnetic radiation from a frequency / wavelength of 3×10^{11} Hz / 1000 μm to 3×10^{16} Hz / 10 nm. The unit used for such measurements is the watt or $m^2 . kg . s^{-3}$.

Photometry is a subset of radiometry in being the measurement of electromagnetic radiation visible to the human eye and weighted by the response of the human eye, which is approximately Gaussian, having a maximum sensitivity around 5.41×10^{14} Hz / 555 nm with cut-offs at 3.86×10^{14} Hz / 770 nm and 7.89×10^{14} Hz / 380 nm (Cox, 2000).

The radiometric unit that corresponds to the SI base unit, the candela, is the **watt per steradian (W . sr^{-1})**. They are related through the formal definition of the candela by:

$$1\,\text{cd} \equiv 1\,\text{lm} . \text{sr}^{-1} \equiv \frac{1}{683}\,\text{W} . \text{sr}^{-1} \tag{8.1}$$

where the ($\frac{1}{683}$) resulted from the need to maintain continuity of units when an earlier definition of the candela was changed.

Given that astronomical observations are now made throughout the electromagnetic spectrum, it would seem appropriate to link astronomical photometry with

Table 8.1. *Photometric, radiometric and astronomical units*

Quantity	Photometry symbol	Radiometry symbol	Astronomy symbol
Power	lumen	total radiant flux	total luminosity
	lm	W	m_{bol}
		spectral radiant flux	monochromatic luminosity
		W . Hz^{-1}	m_ν
Power	lux	irradiance	flux density
per unit area	lm . m^{-2}	W . m^{-2}	W . m^{-2}
		spectral irradiance	monochromatic flux density
		W . m^{-2} . Hz^{-1}	Jy
Power	candela	radiant intensity	
per unit	cd or lm . sr^{-1}	W . sr^{-1}	W . sr^{-1}
solid angle		spectral radiant intensity	
		W . sr^{-1} . Hz^{-1}	W . sr^{-1} . Hz^{-1}
Power	nit	radiance	intensity
per unit area	cd . m^{-2}	W .m^{-2} . sr^{-1}	W .m^{-2} . sr^{-1}
per unit	lm . m^{-2} . sr^{-1}	spectral radiance	specific intensity
solid angle		W . m^{-2} . sr^{-1} . Hz^{-1}	W . m^{-2} . sr^{-1} . Hz^{-1}

radiometry rather than photometry as defined (it should be noted that astronomical observations cover an even larger total bandwidth than that of radiometry).

Equivalent photometric, radiometric and astronomical units and the quantities they represent are set out in Table 8.1.

The jansky is a non-SI unit that is recognized by the IAU for use in astronomy. Formally:

The dimension of **monochromatic flux density or spectral irradiance** is $[\boldsymbol{M}]\,.\,[\boldsymbol{T}]^{-2}$, its unit is the **jansky** with symbol, **Jy**, or **10^{-26} W . m^{-2} . Hz^{-1}**.

The jansky was introduced by radio astronomers and is also used by infrared astronomers. Absolute calibrations of the many magnitude systems (see below) in common use often use the jansky or a similar SI-based or cgs-based unit. Other than the generally very small measurements encountered in astronomy for monochromatic flux density there would seem to be no good reason for not adopting the SI derived radiometric unit for spectral irradiance, the W . m^{-2} . Hz^{-1}. As an example, the monochromatic flux density of a zero-magnitude A0V star in the V waveband according to Bessell (2001) is 3636 Jy. Using the radiometric unit for spectral irradiance, this is equivalent to 3.636×10^{-23} W . m^{-2} . Hz^{-1} or 3.636 d −23 W . m^{-2} . Hz^{-1}.

Monochromatic flux density and total luminosity are the quantities most commonly used in astronomy. An in-depth description of these and other radiometric, photometric and astrophysical terms is given by Sterken & Manfroid (1992).

Example: converting radiometric to photometric units

According to Sackman *et al.* (1993), the value for the solar constant (the total solar irradiance just outside the Earth's atmosphere) $f_\odot$, determined from the combined observations made by several spacecraft, is:

$$f_\odot = 1370 \pm 2\,\mathrm{W\,.\,m^{-2}} \tag{8.2}$$

The surface area of a sphere with a radius r, equal to the mean Sun–Earth distance, is:

$$4\pi r^2 = 4\pi\,(1.4960 \times 10^{11})^2 = 2.8124 \times 10^{23}\,\mathrm{m^2} \tag{8.3}$$

The total solar radiant flux falling on this sphere is:

$$4\pi r^2\,.\,f_\odot = 3.8530 \times 10^{26}\,\mathrm{W} \tag{8.4}$$

(equivalent to a radiant intensity of $r^2\,.\,f_\odot = 3.0661 \times 10^{25}\,\mathrm{W\,.\,sr^{-1}}$). To convert from the radiant flux in watts to the luminous flux in lumens, simply multiply by 683 (the number of lumens in a watt):

$$683 \times 4\pi r^2\,.\,f_\odot = 2.6316 \times 10^{29}\,\mathrm{lm} \tag{8.5}$$

The luminous intensity $I_\odot$, measured in candelas (the SI base unit), is the luminous flux $F_\odot$ per steradian:

$$I_\odot = \frac{F_\odot}{4\pi} = 2.0941 \times 10^{28}\,\mathrm{cd} \tag{8.6}$$

Note that as the Sun is obviously not a point source at the distance of the Earth and does not radiate uniformly in all directions nor uniformly as a function of time, the value calculated for the luminous flux is only an approximation.

Example: stellar magnitude to candelas

Consider an A0V star of apparent magnitude $m_\mathrm{V} = 0.0$ at a distance of 10 pc (3.086×10^{17} m) so that its absolute magnitude $M_\mathrm{V} = 0.0$ as well.

The surface area A of a sphere of radius 10 pc in units of square metres is:

$$A = 4\pi r^2 = 4\pi\,(3.086 \times 10^{17})^2 = 1.1967 \times 10^{36}\,\mathrm{m^2} \tag{8.7}$$

The empirical conversion factor from V magnitudes to janskys is (Bessell, 2001):

$$f_{\mathrm{V}=0} = 3636\,\mathrm{Jy} = 3.636 \times 10^{-23}\,\mathrm{W\,.\,m^{-2}\,.\,Hz^{-1}} \tag{8.8}$$

Total monochromatic flux from the star is:

$$A \,.\, f_{V=0} = 3.636 \times 10^{-23} \times 1.1967 \times 10^{36}$$
$$= 4.3512 \times 10^{13}\,\mathrm{W \,.\, Hz^{-1}} \qquad (8.9)$$

Multiply by the V bandwidth Δ_ν, in frequency units (Dodd, 2007):

$$\Delta_\nu = 8.94 \times 10^{13}\,\mathrm{Hz} \qquad (8.10)$$

to give:

$$A \,.\, \Delta_\nu \,.\, f_{V=0} = 3.8516 \times 10^{27}\,\mathrm{W} \qquad (8.11)$$

In candelas:

$$I_{V=0} = \frac{3.8516 \times 10^{27} \times 683}{4\pi} = 2.1143 \times 10^{29}\,\mathrm{cd} \qquad (8.12)$$

8.2.1 Common astronomical photometric units

It is perhaps a trifle misleading to use the adjective 'common' in the title of this section as on occasion the unit is only ever used by the author of the paper or catalogue to which reference is made. Some examples of the many different units used are:

(i) for wavelengths: Å (angstrom unit $= 10^{-10}$ m), 0.1 nm, nm, μm, mm, cm, and m;
(ii) for frequency: Hz, kHz, MHz, GHz, THz, and PHz;
(iii) for monochromatic flux densities: $\mathrm{erg \,.\, s^{-1} \,.\, cm^{-2}}$; $\mathrm{erg \,.\, s^{-1} \,.\, Å^{-1}}$; $\mathrm{erg \,.\, s^{-1}}$; magnitudes (many different systems); Jy; $\mathrm{erg \,.\, Hz^{-1} \,.\, s^{-1}}$; $\mathrm{mW \,.\, m^{-2} \,.\, (0.1\,nm)^{-1}}$.

The many attempts to produce conversion factors from magnitudes to flux densities have not been of great assistance either, as the quoted flux values generally have their own units as well, e.g.:

UBV photometry: for a V = 0, A0V star, $f_V = 3636 \times 10^{-30}\,\mathrm{W\,.cm^{-2}\,Hz^{-1}}$
Vilnius photometry for a U = 0, OV star, $f_U = 19.22 \times 10^{-12}\,\mathrm{W\,cm^{-2}\,\mu m^{-1}}$
2MASS infrared photometry for a J = 0 star, $f_J = 1592 \pm 15.2\,\mathrm{Jy}$

Given the plethora of units in use at present, astronomers must surely benefit by adopting a single set to be used by all. In this book, SI photometric units are abandoned in favour of SI derived radiometric units, as well as substituting these units for magnitudes. Since the IAU approves the use of the jansky, which is essentially a prefixed (by 10^{-26}) SI derived unit ($\mathrm{W \,.\, m^{-2} \,.\, Hz^{-1}}$), all transformations from other flux units and from magnitudes are made to janskys. A substitution for Hipparchus/Pogson magnitudes using a 'jansky' magnitude $\ln(f_\nu)$, where f_ν is the monochromatic flux in janskys, is also introduced.

8.2.2 Wavelengths to frequencies

Consider a device (e.g., a glass filter in an optical telescope) that isolates a portion of the electromagnetic spectrum of bandwidth $\Delta\lambda$ (taken to mean the full width at half maximum height (FWHM) centred on a wavelength λ). The following equation is used to compute the frequency ν, equivalent to the wavelength λ, where c is the velocity of light (=299 792 458 m . s^{-1}) in vacuo:

$$\nu = \frac{c}{\lambda} \tag{8.13}$$

A filter with a bandwidth $\Delta\lambda$, defined by the FWHM, symmetrically placed about the mean wavelength λ, does not have a symmetrical bandpass in frequency space about the frequency ν, corresponding to the wavelength λ. To compute the bandwidth and locate the upper and lower frequency bounds the following method may be used:

$$\nu_{\mathrm{hfb}} = \frac{c}{\lambda - 0.5\,\Delta\lambda} \tag{8.14}$$

$$\nu_{\mathrm{lfb}} = \frac{c}{\lambda + 0.5\,\Delta\lambda} \tag{8.15}$$

$$\Delta\nu = \nu_{\mathrm{hfb}} - \nu_{\mathrm{lfb}} \tag{8.16}$$

where ν_{hfb} is the cut-off frequency at the high-frequency boundary of the FWHM location of the filter and ν_{lfb} is the cut-off frequency at the low-frequency boundary of the FWHM location of the filter. Figure 8.2 illustrates the differences between bandwidths in wavelength and frequency spaces.

Example: convert from wavelength to frequency space for the Johnson V band

Bessell (2001) gives the effective wavelength and FWHM bandwidth of the Johnson–Cousins–Glass V band as: $\lambda = 545$ nm and $\Delta\lambda = 85$ nm. In metres, these values become 5.45×10^{-7} and 8.8×10^{-8}. Substituting for λ in Equation (8.13) gives:

$$\nu = \frac{299792458}{5.45\times10^{-7}} = 5.501\times10^{14}\,\mathrm{Hz} \tag{8.17}$$

The short-wavelength end of the V-filter FWHM is at:

$$\lambda - \frac{1}{2}\Delta\lambda = 5.025\times10^{-7}\,\mathrm{m} \tag{8.18}$$

and the long-wavelength end is at:

$$\lambda + \frac{1}{2}\Delta\lambda = 5.875\times10^{-7}\,\mathrm{m} \tag{8.19}$$

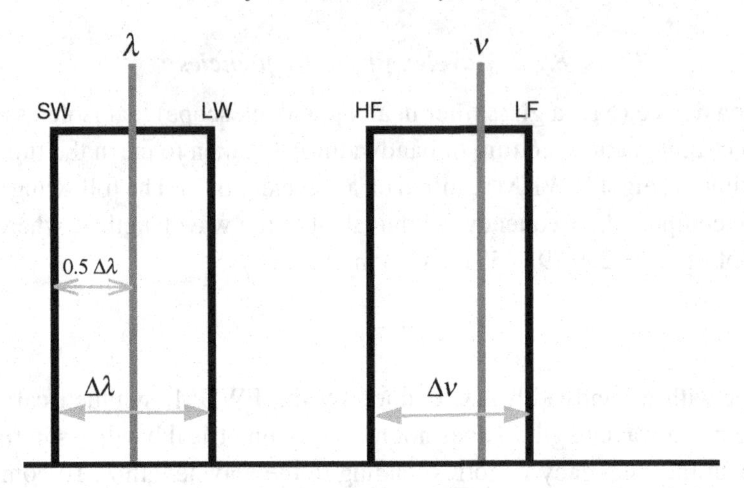

Figure 8.1. Bandwidths in wavelength and frequency spaces. SW is short wavelength, LW is long wavelength; HF is high frequency and LF is low frequency.

Compute the corresponding frequencies ν_{hfb} and ν_{lfb} using Equations (8.14) and (8.15) to give $\nu_{\text{hfb}} = 5.966\,0 \times 10^{14}$ Hz and $\nu_{\text{lfb}} = 5.102\,9 \times 10^{14}$ Hz. The frequency bandwidth is simply the difference between the two:

$$\Delta\nu = 8.631 \times 10^{13}\,\text{Hz}. \tag{8.20}$$

Note that the wavelength FWHM is symmetrically placed with respect to the mean wavelength but the frequency FWHM is not (see Figure 8.2).

8.2.3 The conversion of some commonly used photometric systems from wavelength to frequency space

Using the method set out in the worked example above, effective wavelengths and bandwidths in nanometres were converted into frequencies in hertz and listed in Table 8.2. Also included in the table are the low- and high-frequency bounds of the FWHM of each filter (or isolated spectral band) in hertz.

Wavelength data for the Johnson–Cousins–Glass photometry, Geneva photometry, Strömgren photometry, Walraven photometry, DDO photometry, Washington photometry, Thuan–Gunn photometry (all photoelectric systems); SDSS (Sloan Digital Sky Survey) photometry, UBV photometry, MACHO and EROS photometry (gravitational microlensing observational programmes), which are all ground-based CCD photometric systems; HST and HIPPARCOS/TYCHO photometry (space-based photometric systems) were obtained from Bessell (2001).

The photographic I-band photometry of the USNO-B astrometric catalogue was derived by Dodd (2007). The Spitzer infrared space telescope filter specifications are given in the online MIPS (Multiband Imaging Photometer for

Table 8.2. *Conversion of effective wavelengths and bandwidths from units of length to frequency units*

Filter	λ (nm) $\Delta\lambda$ (nm)	ν (Hz) $\Delta\nu$ (Hz)	ν_{lfb} (Hz)	ν_{hfb} (Hz)
	Johnson	Cousins	Glass	
U	367	8.169 d 14	7.495 d 14	8.976 d 14
	66	1.481 d 14		
B	436	6.876 d 14	6.207 d 14	7.707 d 14
	94	1.500 d 14		
V	545	5.501 d 14	5.103 d 14	5.966 d 14
	85	8.632 d 13		
R	638	4.699 d 14	4.175 d 14	5.373 d 14
	160	1.197 d 14		
I	797	3.762 d 14	3.440 d 14	4.149 d 14
	149	7.094 d 13		
J	1220	2.457 d 14	2.260 d 14	2.692 d 14
	213	4.323 d 13		
H	1630	1.839 d 14	1.681 d 14	2.030 d 14
	307	3.495 d 13		
K	2190	1.369 d 14	1.257 d 14	1.503 d 14
	390	2.457 d 13		
L	3450	8.690 d 13	8.133 d 13	9.328 d 13
	472	1.194 d 13		
M	4750	6.311 d 13	6.020 d 13	6.633 d 13
	460	6.126 d 12		
		Geneva		
U	350	8.565 d 14	8.027 d 14	9.182 d 14
	47	1.155 d 14		
B	424	7.071 d 14	6.489 d 14	7.767 d 14
	76	1.278 d 14		
B_1	402	7.458 d 14	7.121 d 14	7.827 d 14
	38	7.065 d 13		
B_2	448	6.692 d 14	6.399 d 14	7.013 d 14
	41	6.137 d 13		
V	551	5.441 d 14	5.129 d 14	5.793 d 14
	67	6.640 d 13		
V_1	541	5.541 d 14	5.325 d 14	5.776 d 14
	44	4.514 d 13		
G	578	5.187 d 14	4.984 d 14	5.407 d 14
	47	4.225 d 13		
		Strömgren		
u	349	8.590 d 14	8.236 d 14	8.976 d 14
	30	7.398 d 13		
v	411	7.294 d 14	7.129 d 14	7.467 d 14
	19	3.374 d 13		
b	467	6.420 d 14	6.298 d 14	6.546 d 14
	18	2.475 d 13		

Table 8.2. (*cont.*)

Filter	λ (nm) $\Delta\lambda$ (nm)	ν (Hz) $\Delta\nu$ (Hz)	$\nu_{\rm lfb}$ (Hz)	$\nu_{\rm hfb}$ (Hz)
y	547	5.481 d 14	5.368 d 14	5.598 d 14
	23	2.306 d 13		
β_w	489	6.131 d 14	6.038 d 14	6.226 d 14
	15	1.881 d 13		
β_n	486	6.169 d 14	6.150 d 14	6.188 d 14
	3	3.808 d 12		
		Walraven		
W	323.3	9.273 d 14	9.057 d 14	9.499 d 14
	15.4	4.420 d 13		
U	361.6	8.291 d 14	8.037 d 14	8.561 d 14
	22.8	5.233 d 13		
L	383.5	7.817 d 14	7.600 d 14	8.047 d 14
	21.9	4.468 d 13		
B	427.7	7.009 d 14	6.630 d 14	7.435 d 14
	49.0	8.057 d 13		
V	540.6	5.546 d 14	5.207 d 14	5.931 d 14
	70.3	7.242 d 13		
		DDO		
35	349.0	8.590 d 14	8.143 d 14	9.089 d 14
	38.3	9.455 d 13		
38	381.5	7.858 d 14	7.532 d 14	8.213 d 14
	33.0	6.810 d 13		
41	416.6	7.196 d 14	7.125 d 14	7.269 d 14
	8.3	1.434 d 13		
42	425.7	7.042 d 14	6.982 d 14	7.103 d 14
	7.3	1.208 d 13		
45	451.7	6.637 d 14	6.582 d 14	6.693 d 14
	7.6	1.117 d 13		
48	488.6	6.136 d 14	6.021 d 14	6.255 d 14
	18.6	2.337 d 13		
	Thuan	Gunn		
u	353	8.493 d 14	8.037 d 14	9.003 d 14
	40	9.654 d 13		
v	398	7.532 d 14	7.172 d 14	7.931 d 14
	40	7.589 d 13		
g	493	6.081 d 14	5.678 d 14	6.546 d 14
	70	8.678 d 13		
r	655	4.577 d 14	4.283 d 14	4.915 d 14
	90	6.319 d 13		
		Vilnius		
U	345	8.690 d 14	8.213 d 14	9.224 d 14
	40	1.011 d 14		
P	374	8.016 d 14	7.747 d 14	8.305 d 14
	26	5.579 d 13		

Table 8.2. (*cont.*)

Filter	λ (nm) $\Delta\lambda$ (nm)	ν (Hz) $\Delta\nu$ (Hz)	ν_{lfb} (Hz)	ν_{hfb} (Hz)
X	405	7.402 d 14	7.207 d 14	7.609 d 14
	22	4.024 d 13		
Y	466	6.433 d 14	6.259 d 14	6.618 d 14
	26	3.592 d 13		
Z	516	5.810 d 14	5.694 d 14	5.931 d 14
	21	2.365 d 13		
V	544	5.511 d 14	5.382 d 14	5.646 d 14
	26	2.635 d 13		
S	656	4.570 d 14	4.501 d 14	4.641 d 14
	20	1.394 d 13		
		2MASS		
J	1235	2.427 d 14	2.278 d 14	2.598 d 14
	162	3.198 d 13		
H	1662	1.804 d 14	1.677 d 14	1.951 d 14
	251	2.740 d 13		
K	2159	1.389 d 14	1.309 d 14	1.478 d 14
	261	1.685 d 13		
	SDSS	A0 star		
u′	356	8.421 d 14	7.727 d 14	9.253 d 14
	64	1.526 d 14		
g′	475	6.311 d 14	5.526 d 14	7.357 d 14
	135	1.831 d 14		
r′	620	4.835 d 14	4.354 d 14	5.436 d 14
	137	1.082 d 14		
i′	761	3.939 d 14	3.577 d 14	4.383 d 14
	154	8.055 d 13		
z′	907	3.305 d 14	3.058 d 14	3.597 d 14
	147	5.392 d 13		
		Washington		
C	391	7.667 d 14	6.722 d 14	8.922 d 14
	110	2.201 d 14		
M	509	5.890 d 14	5.339 d 14	6.567 d 14
	105	1.228 d 14		
T_1	633	4.736 d 14	4.455 d 14	5.056 d 14
	80	6.010 d 13		
T_2	805	3.724 d 14	3.407 d 14	4.107 d 14
	150	7.000 d 13		
		ANS		
ANS1	154.5	1.940 d 15	1.910 d 15	1.972 d 15
	5.0	6.281 d 13		
ANS2	154.9	1.935 d 15	1.847 d 15	2.033 d 15
	14.9	1.866 d 14		
ANS3	179.9	1.666 d 15	1.600 d 15	1.738 d 15
	14.9	1.383 d 14		

Table 8.2. (*cont.*)

Filter	λ (nm) $\Delta\lambda$ (nm)	ν (Hz) $\Delta\nu$ (Hz)	ν_{lfb} (Hz)	ν_{hfb} (Hz)
ANS4	220.0	1.363 d 15	1.303 d 15	1.428 d 15
	20.0	1.241 d 14		
ANS5	249.3	1.203 d 15	1.167 d 15	1.240 d 15
	15.0	7.242 d 13		
ANS6	329.4	9.101 d 14	8.965 d 14	9.241 d 14
	10.0	2.764 d 13		
		TD1		
TD1	156.5	1.916 d 15	1.733 d 15	2.141 d 15
	33	4.085 d 14		
TD2	196.5	1.526 d 15	1.407 d 15	1.666 d 15
	33	2.580 d 14		
TD3	236.5	1.268 d 15	1.185 d 15	1.363 d 15
	33	1.777 d 14		
TD4	274.0	1.094 d 15	1.036 d 15	1.160 d 15
	31	1.242 d 14		
		IRAS		
12 μm	12 000	2.498 d 13	1.934 d 13	3.527 d 13
	7 000	1.593 d 13		
25 μm	25 000	1.199 d 13	9.805 d 12	1.543 d 13
	11 150	5.628 d 12		
60 μm	60 000	4.997 d 12	3.932 d 12	6.852 d 12
	32 500	2.921 d 12		
100 μm	100 000	2.998 d 12	2.590 d 12	3.558 d 12
	31 500	9.684 d 11		
		HST		
HST336	334	8.976 d 14	8.386 d 14	9.655 d 14
	47	1.269 d 14		
HST439	430	6.972 d 14	6.440 d 14	7.599 d 14
	71	1.159 d 14		
HST450	451	6.647 d 14	5.942 d 14	7.542 d 14
	107	1.600 d 14		
HST555	532	5.635 d 14	4.951 d 14	6.539 d 14
	147	1.587 d 14		
HST675	667	4.495 d 14	4.104 d 14	4.968 d 14
	127	8.636 d 13		
HST814	788	3.804 d 14	3.480 d 14	4.196 d 14
	147	7.159 d 13		
	HIPPARCOS	TYCHO		
B_T	421	7.121 d 14	6.574 d 14	7.767 d 14
	70	1.192 d 14		
V_T	526	5.699 d 14	5.205 d 14	6.298 d 14
	100	1.093 d 14		
H_P	517	5.799 d 14	4.744 d 14	7.458 d 14
	230	2.714 d 14		

Table 8.2. (*cont.*)

Filter	λ (nm) Δλ (nm)	ν (Hz) Δν (Hz)	$\nu_{\rm lfb}$ (Hz)	$\nu_{\rm hfb}$ (Hz)
		MACHO		
B	519	5.776 d 14	5.073 d 14	6.707 d 14
	144	1.634 d 14		
R	682	4.396 d 14	3.888 d 14	5.056 d 14
	178	1.167 d 14		
		UBV(CCD)		
B	436	6.876 d 14	6.207 d 14	7.707 d 14
	94	1.500 d 14		
V	545	5.501 d 14	5.103 d 14	5.966 d 14
	85	8.632 d 13		
R	641	4.677 d 14	4.158 d 14	5.344 d 14
	160	1.186 d 14		
I	791	3.790 d 14	3.476 d 14	4.167 d 14
	143	6.908 d 13		
Z	909	3.298 d 14	3.133 d 14	3.482 d 14
	96	3.493 d 13		
		USNO-B(I)		
I	807.5	3.713 d 14	3.331 d 14	4.193 d 14
	185	8.619 d 13		
		EROS		
BE1	485	6.181 d 14	5.557 d 14	6.964 d 14
	109	1.407 d 14		
BE2	539	5.562 d 14	4.729 d 14	6.752 d 14
	190	2.023 d 14		
RE1	657	4.563 d 14	3.984 d 14	5.339 d 14
	191	1.355 d 14		
RE2	767	3.909 d 14	3.342 d 14	4.706 d 14
	260	1.364 d 14		
		Spitzer		
24 μm	23700	1.265 d 13	1.151 d 13	1.404 d 13
	4700	2.533 d 12		
70 μm	71000	4.222 d 12	3.724 d 12	4.875 d 12
	19000	1.151 d 12		
160 μm	156000	1.922 d 12	1.728 d 12	2.165 d 12
	35000	4.367 d 11		
		DENIS		
i	791	3.790 d 14	3.476 d 14	4.167 d 14
	143	6.910 d 13		
J	1228	2.441 d 14	2.265 d 14	2.647 d 14
	191	3.820 d 13		
K_S	2145	1.398 d 14	1.306 d 14	1.503 d 14
	302	1.970 d 13		

In this table the shorthand notation $m\,{\rm d}\,n$ has been used where $m\,{\rm d}\,n \equiv m \times 10^n$.

Spitzer) instrument handbook.[49] The ultraviolet photometer and filters used by the ANS (Astronomical Netherlands Satellite) are described by van Duinen *et al.* (1975). A description of the equipment used by the TD1 photometric satellite is given in the introduction of the *Catalogue of Ultraviolet Stellar Fluxes* by Thompson *et al.* (1978).

Detailed information on the Vilnius seven-colour photometry was obtained from Straižys (1992). The absolute calibration of the 2MASS infrared photometric system and the filter parameters are given by Cohen *et al.* (2003), and finally data on the Infrared Astronomical Satellite (IRAS) is given on its website.[50]

8.2.4 Determination of the spectral irradiance and the monochromatic flux density of stars from a selection of published values in non-SI units

In the introduction to the third edition of a *Catalog of Infrared Observations*, Gezari *et al.* (1993) list 26 different ways in which the infrared stellar monochromatic flux density of stars are given in the catalogue. Some of the units are cgs based, some are SI based, some in wavelength space, some in frequency space, some are magnitudes, some magnitudes per square arcsecond and some are derived from spectrophotometric measures. Inter-comparison between the different units is not easy and often prone to mistakes.

Various specific examples are set out below, beginning with monochromatic flux densities given in non-SI and ending with the conversion of the most common unit of stellar brightness unit, the magnitude, into janskys and the radiometric spectral irradiance unit $W \,.\, m^{-2} \,.\, Hz^{-1}$.

Conversion of apparent monochromatic flux density from wavelength space to frequency space

Many of the earlier monochromatic flux density units are in wavelength space and so must be converted into frequency space to produce measures in janskys or $W \,.\, m^{-2} \,.\, Hz^{-1}$. If f_λ is the apparent monochromatic flux density in a wavelength band $\Delta\lambda$ wide, centred on wavelength λ, both measured in metres, and $\Delta\nu$ is the bandwidth in hertz, then f_ν, the apparent monochromatic flux density in $W \,.\, m^{-2} \,.\, Hz^{-1}$ at frequency ν (derived from Equation 8.13), is given by:

$$f_\nu = f_\lambda \cdot \frac{\Delta\lambda}{\Delta\nu} \tag{8.21}$$

[49] See http://www.irsa.ipac.caltech.edu/data/SPITZER/docs/mips/mipsinstrumenthandbook/
[50] See http://irsa.ipac.caltech.edu/IRASdocs/

where $\Delta\nu$ is determined using Equation (8.16). In janskys, the expression is simply:

$$f_\nu = f_\lambda . \frac{\Delta\lambda}{\Delta\nu} . 10^{26} \qquad (8.22)$$

Conversion of magnitudes to monochromatic flux density

The majority of catalogues of the brightness of stars from the ultraviolet to the near infrared use magnitudes as a measure of that brightness. The scale of magnitude is logarithmic, with a base of $2.512 = 10^{0.4}$ (i.e., the fifth root of 100). The magnitude system is ordinal, so that the brighter the object the smaller the magnitude: first magnitude objects are brighter than second magnitude objects and so on. The system was devised by the Greek astronomer Hipparchus over 2000 years ago, and made more respectable by the English astronomer Pogson in the nineteenth century when he fixed the brightness ratio of 5 magnitudes to equal a factor of 100.

The apparent magnitude, m_λ, of a celestial body at effective wavelength λ, through a filter of bandwidth $\Delta\lambda$, may be defined in terms of the monochromatic flux density as:

$$m_\lambda = -2.5 \log f_\lambda \qquad (8.23)$$

If this Equation (8.23) is inverted and converted to a natural logarithm base, rather than base 10, then:

$$f_\lambda = f_{(m_\lambda=0)}\, \mathrm{e}^{-0.921\, m_\lambda} \qquad (8.24)$$

where $f_{m_\lambda=0}$ is the measured monochromatic flux density in a bandwidth centred on wavelength λ of a star of magnitude m_λ =0.

Conversions of some published non-SI flux units to janskys

Example 1

Bessell (2001) gives the absolute calibration of the Johnson–Cousins–Glass broadband photometry in units of $\mathbf{10^{-30} . W . cm^{-2} . Hz^{-1}}$. Now

$$1\,\mathrm{Jy} = 10^{-26}\,\mathrm{W.m^{-2}.Hz^{-1}}$$

$$\begin{aligned} 10^{-30}\,\mathrm{W.cm^{-2}.Hz^{-1}} &= 10^{-30}\,\mathrm{W.(0.01.m)^{-2}.Hz^{-1}} \\ &= 10^{-30}.10^{4}.\mathrm{W.m^{-2}.Hz^{-1}} \\ &= 10^{-26}\,\mathrm{W.m^{-2}.Hz^{-1}} \\ &= 1\,\mathrm{Jy} \end{aligned} \qquad (8.25)$$

So the units used by Bessell were essentially janskys expressed in a hybrid cgs and SI system.

Example 2

Convert the Vilnius U-band monochromatic flux density $f_{\lambda(U=0)}$ (Straižys, 1992) expressed in units of $\mathbf{W.cm^{-2}.\mu m^{-1}}$ into janskys, for a hypothetical OV-type star.

$$\begin{aligned} f_{\lambda(\mathrm{U}=0)} &= 19.22 \times 10^{-12}\,\mathrm{W.cm^{-2}.\mu m^{-1}} \\ &= 19.22 \times 10^{-12}\,\mathrm{W.(10^{-2}.m)^{-2}.(10^{-6}.m)^{-1}} \\ &= 19.22 \times 10^{-12} \times 10^4 \times 10^6\,\mathrm{W.m^{-2}.m^{-1}} \\ &= 19.22 \times 10^{-2}\,\mathrm{W.m^{-2}.m^{-1}} \end{aligned} \tag{8.26}$$

From Table 8.2, extract the appropriate values for bandwidth in wavelength ($\Delta\lambda_{\mathrm{U}}$) and frequency ($\Delta\nu_{\mathrm{U}}$) space and substitute these values into Equation (8.22) to determine the value of the monochromatic flux density in janskys:

$$\Delta\lambda_{\mathrm{U}} = 4 \times 10^{-8}\,\mathrm{m}\ \Delta\nu_{\mathrm{U}} = 1.011 \times 10^{14}\,\mathrm{Hz}$$

$$\begin{aligned} f_{\nu(\mathrm{U})} &= 19.22 \times 10^{-2} \,.\, \frac{4 \times 10^{-8}}{1.011 \times 10^{14}} \,.\, 10^{26} \\ &= 7.604 \times 10^3\,\mathrm{Jy} \end{aligned} \tag{8.27}$$

In radiometric units, the spectral irradiance is $7.604\times10^{-23}\,\mathrm{W.m^{-2}.Hz^{-1}}$.

Example 3

Convert the TD1-S68 ultraviolet satellite measure at $\lambda = 274$ nm (Thompson *et al.*, 1978) expressed in $\mathbf{erg.cm^{-2}.s^{-1}.\mathring{A}^{-1}}$ into janskys for the A0V star Vega.

$$\begin{aligned} f_{274} &= 3.123 \times 10^{-9}\,\mathrm{erg.cm^{-2}.s^{-1}.\mathring{A}^{-1}} \\ &= 3.123 \times 10^{-9}\,(10^{-7}\,\mathrm{J}).\mathrm{s^{-1}}.(10^{-2}\,\mathrm{m})^{-2}.(10^{-10}\,\mathrm{m})^{-1} \\ &= 3.123 \times 10^{-9} \times 10^{-7} \times 10^4 \times 10^{10}\,\mathrm{J.s^{-1}.m^{-2}.m^{-1}} \\ &= 3.123 \times 10^{-2}\,\mathrm{W.m^{-2}.m^{-1}} \end{aligned} \tag{8.28}$$

From Table 8.2, $\Delta\lambda_{274} = 3.1 \times 10^{-8}$ and $\Delta\nu_{274} = 1.242\times10^{14}$ Hz, so, using Equation (8.22):

$$\begin{aligned} f_{\nu_{274}} &= 3.123 \times 10^{-2} \,.\, \frac{3.1 \times 10^{-8}}{1.242 \times 10^{14}} \,.\, 10^{26}\,\mathrm{Jy} \\ &= 779.5\,\mathrm{Jy} \end{aligned} \tag{8.29}$$

In radiometric units the spectral irradiance is $7.795\times10^{-24}\,\mathrm{W.m^{-2}.Hz^{-1}}$.

Example 4

HST spectrophotometry of the star Vega from 170 nm to 1010 nm. Bohlin & Gilliland (2004) used observations obtained with the Space Telescope Imaging Spectrograph (STIS) to determine the flux density in units of $\mathrm{mW\,.\,m^{-2}\,.\,(0.1\,nm)^{-1}}$ (where $0.1\,\mathrm{nm} = 1\,\mathrm{\AA}$) against wavelength in angstroms. A plot of their data converted to SI units (i.e., wavelength in metres and monochromatic flux densities in $\mathrm{W\,.\,m^{-2}\,.\,m^{-1}}$) is shown in Figure 8.2.

The conversion from Bohlin & Gilliland (2004) units to janskys is carried out for the flux measurement of $3.04\times10^{-9}\,\mathrm{mW\,.\,m^{-2}\,.\,(0.1\,nm)^{-1}}$ determined at a wavelength of 5800 Å and an FWHM measurement of 11.6 Å:

$$\begin{aligned} f_\lambda &= 3.04\times10^{-9}\,\mathrm{mW\,.\,m^{-2}\,.\,(0.1\,nm)^{-1}} \\ &= 3.04\times10^{-9}\,.\,(10^{-3}\,.\,\mathrm{W})\,.\,\mathrm{m^{-2}}\,.\,(10^{-10}\,.\,\mathrm{m})^{-1} \\ &= 3.04\times10^{-9}\,.\,(10^{-3}\,.\,10^{10})\,.\,\mathrm{W\,.\,m^{-2}\,.\,m^{-1}} \\ &= 3.04\times10^{-2}\,\mathrm{W\,.\,m^{-2}\,.\,m^{-1}} \end{aligned} \tag{8.30}$$

To convert a wavelength-space monochromatic flux density f_λ, at wavelength λ, into a frequency-space monochromatic flux density f_ν, at a frequency ν, corresponding to λ, Equation (8.22) is used. Bohlin & Gilliland (2004) list values of the FWHM bandwidth at each central wavelength sampled, which is taken as $\Delta\lambda$ and $\Delta\nu$ then calculated using Equations (8.13) to (8.16). Finally, a value for the

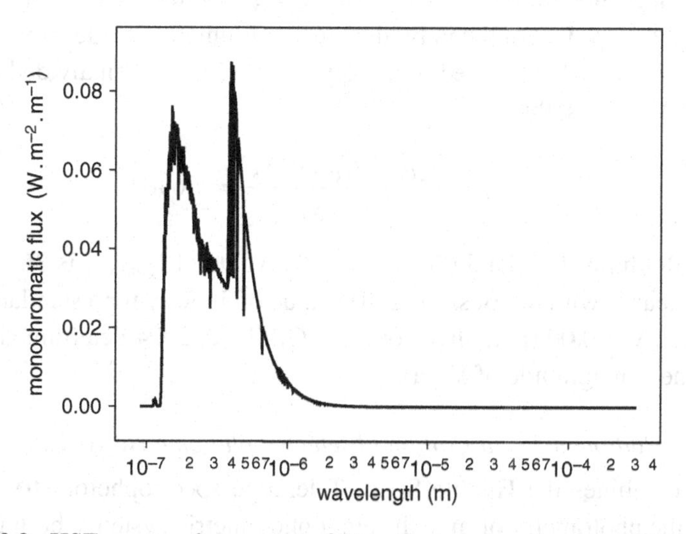

Figure 8.2. HST spectrophotometry of Vega in wavelength space. Wavelength is measured in metres (logarithmic scale) and monochromatic flux density in radiometric units ($\mathrm{W\,.\,m^{-2}\,.\,m^{-1}}$).

frequency monochromatic flux density is derived from Equation (8.22). For the numerical example, entry 2436 in the Bohlin & Gilliland (2004) tabulation is used and firstly converted to SI units:

$$c = 299\,792\,458\,\mathrm{m.s^{-1}}$$
$$\lambda = 5.80 \times 10^{-7}\,\mathrm{m}$$
$$\Delta\lambda = 1.16 \times 10^{-9}\,\mathrm{m} \tag{8.31}$$

Convert from wavelength to frequency:

$$\nu = \frac{c}{\lambda} = 5.169 \times 10^{14}\,\mathrm{Hz}$$
$$\nu_{\mathrm{hfb}} = \frac{c}{\lambda - 0.5\,\Delta\lambda} = 5.174 \times 10^{14}\,\mathrm{Hz}$$
$$\nu_{\mathrm{lfb}} = \frac{c}{\lambda + 0.5\,\Delta\lambda} = 5.164 \times 10^{14}\,\mathrm{Hz}$$
$$\Delta\nu = \nu_{\mathrm{hfb}} - \nu_{\mathrm{lfb}} = 1.034 \times 10^{12}\,\mathrm{Hz}$$
$$f_\nu = f_\lambda \frac{\Delta\lambda}{\Delta\nu}\,.\,10^{26} = 3415\,\mathrm{Jy} \tag{8.32}$$

As an additional example that utilizes the Bohlin & Gilliland (2004) data, a subset was extracted bounded by the bandwidth of the Johnson V filter. In total, some 157 data points are available within the V bandwidth in the HST Vega data. As above, $(\lambda, \Delta\lambda, f_\lambda)$ were converted to $(\nu, \Delta\nu, f_\nu)$ and a plot drawn (Figure 8.3) of ν against f_ν. The mean monochromatic flux density $f_{(\nu=\mathrm{V}_\nu)}$, within the V frequency band, is the sum of the product of the individual monochromatic flux densities multiplied by the FWHM of each measured frequency sample $(\nu\Delta\nu)$ then divided by the sum of the FWHM values, thus:

$$f_{(\nu=\mathrm{V}_\nu)} = \frac{\sum_{k=1}^{157}(f_{\nu_k})\,.\,(\Delta\nu_k)}{\sum_{k=1}^{157}(\Delta\nu_k)} = 3646\,\mathrm{Jy} \tag{8.33}$$

For the Bohlin & Gilliland (2004) data, the value of $\mathrm{f}_{(\nu=\mathrm{V}_\nu)}$ is 3646 Jy, which may be compared with the Bessell (2001) value of 3636 Jy for a standard A0V star of magnitude V = 0.00 (note that Bohlin & Gilliland, 2004 determined a value of 0.026 for the V magnitude of Vega).

Radiometric calibration of selected photometric systems

Figure 8.5 combines the Hubble Space Telescope spectrophotometry of the star Vega with the photometry of nine different photometric systems, both ground and space based, converted in each case to janskys. The spectral irradiance in the radiometric unit $\mathrm{W\,.\,m^{-2}\,.\,Hz^{-1}}$ is obtained from the monochromatic flux density in

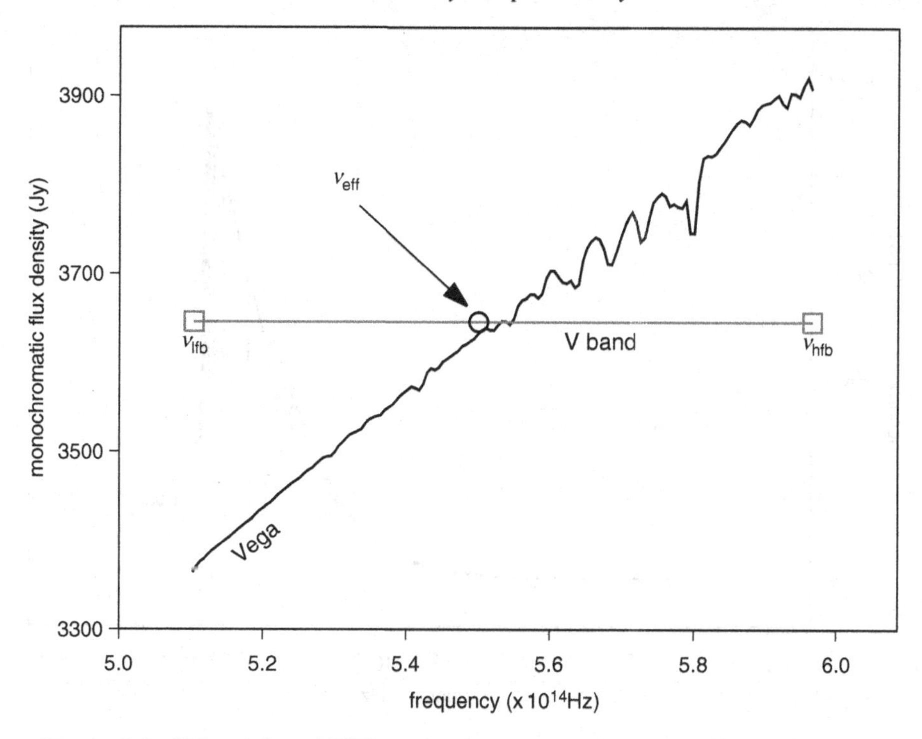

Figure 8.3. V band from HST spectrophotometry of Vega in frequency space. Frequency is measured in hertz and monochromatic flux density in janskys (10^{-26} . W . m^{-2} . Hz^{-1}), ν_{eff} is the frequency equivalent to the effective wavelength λ_{eff}, the squares mark the extreme ends of the FWHM points of the bandwidth of the Johnson V filter and are labelled ν_{lfb} for the low-frequency bound and ν_{hfb} for the high-frequency bound.

janskys by multiplying by 10^{-26}. In all, some 47 individual photometric points are plotted with their frequency FWHM bandwidth shown as a horizontal error bar. Some of the monochromatic flux densities were taken directly from published values, some by converting cgs or hybrid units into janskys and some by transforming system magnitudes into janskys. Notes on the methods used in each case are given below. Table 8.3 sets out for each of the identified filters, the filter set to which it belongs and its identifier within that set, the effective frequency in hertz, the frequency bandwidth (FWHM) in hertz and the monochromatic flux density in janskys. The table is arranged in order of increasing frequency, beginning in the infrared and ending in the ultraviolet part of the spectrum.

8.2.5 Low-frequency, high-frequency and wavenumber unit conversions

Low frequency: The low-frequency part of the electromagnetic spectrum is the domain of radio astronomy that covers frequencies from less than 30 MHz to

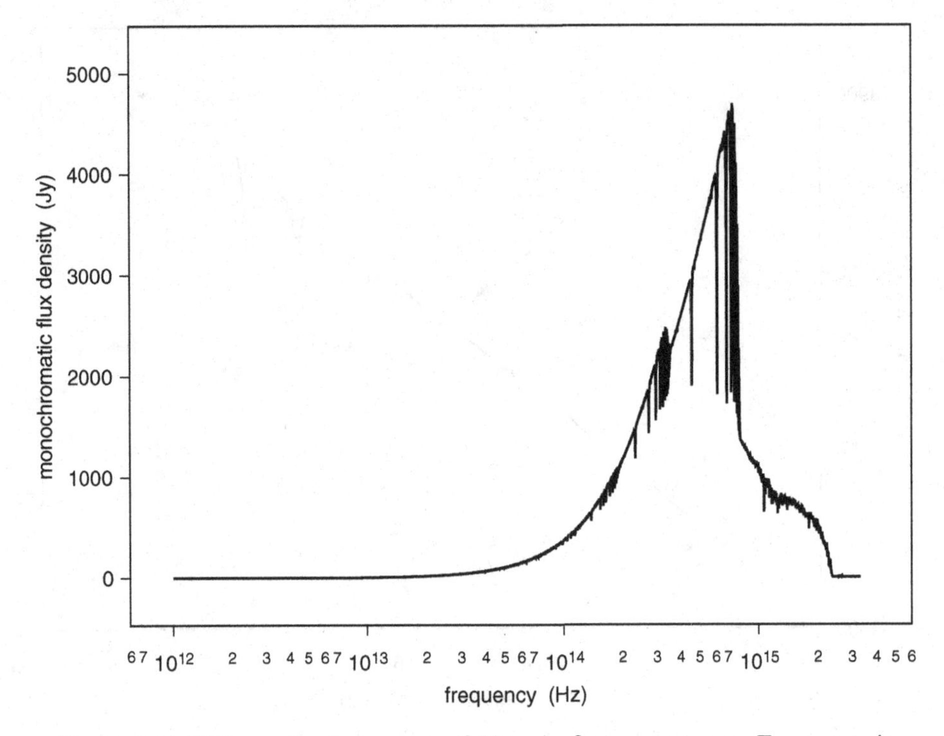

Figure 8.4. HST spectrophotometry of Vega in frequency space. Frequency is measured in hertz (logarithmic scale) and monochromatic flux density in janskys (10^{-26} . W . m^{-2} . m^{-1}). Note that in this figure the photon energy increases from left to right in contrast with the wavelength figure, where the highest-energy photons are at the left-hand side of the plot.

30 GHz over which ground-based observations may be made. Astronomical sources cannot be observed from the ground at frequencies much below 30 MHz due to the reflective properties of the ionosphere. At frequencies greater than 30 GHz, molecular absorption by the lower atmosphere restricts observations to discrete windows. The wavelength range corresponding to these frequencies is approximately 1 cm to a few tens of metres.

In comparison with optical astronomy, radio astronomy is a relatively new science and was established by scientists and engineers with expertise in fields such as electrical engineering and physics rather than astronomy. As such it developed independently its own units which, fortunately, have tended to be based on the physics of the day.

To specify the part of the electromagnetic spectrum that is being observed, radio astronomers use frequency measured in Hz.

The unit of spectral irradiance or monochromatic flux density (termed specific flux or flux density by radio astronomers (Burke & Graham-Smith, 2002)) is the

Table 8.3. *Monochromatic flux densities, either measured observationally for the A0V star Vega or determined by other means for a standard A0V star*

Filter identifier	Effective frequency (Hz)	Frequency bandwidth (Hz)	MFD FWHM (Jy)
iras 100 μm	2.998d12	9.684d11	0.39
iras 60 μm	4.997d12	2.921d12	1.09
iras 25 μm	1.199d13	5.625d12	6.29
iras 12 μm	2.498d13	1.593d13	26.97
jcg M	6.311d13	6.162d12	154.00
jcg L	8.690d13	1.194d13	285.00
jcg K	1.369d14	2.457d13	640.00
2mass K_S	1.389d14	1.690d13	666.70
den K_S	1.398d14	1.970d13	665.00
2mass H	1.804d14	2.740d13	1024.00
jcg H	1.839d14	3.495d13	1020.00
2mass J	2.427d14	3.200d13	1594.00
den J	2.441d14	3.820d13	1595.00
jcg J	2.457d14	4.323d13	1589.00
jcg I	3.762d14	7.094d13	2416.00
den i	3.790d14	6.910d13	2499.00
vil S	4.570d14	1.394d13	3142.04
jcg R	4.699d14	1.197d14	3064.00
gen G	5.187d14	4.225d13	3321.24
gen V	5.441d14	6.640d13	3588.39
strom y	5.481d14	2.306d13	3690.37
jcg V	5.501d14	8.632d13	3636.00
vil V	5.511d14	2.635d13	3749.53
gen V1	5.541d14	4.514d13	3548.95
vil Z	5.810d14	2.365d13	3951.37
strom v	6.420d14	2.475d13	4225.45
vil Y	6.433d14	3.592d13	4169.27
gen B2	6.692d14	6.137d13	4131.43
jcg B	6.876d14	1.500d14	4063.00
gen B	7.071d14	1.278d14	3634.96
strom b	7.294d14	3.374d13	4043.27
vil X	7.402d14	4.023d13	4105.86
gen B1	7.458d14	7.065d13	3927.35
vil P	8.016d14	5.579d13	2111.13
jcg U	8.169d14	1.481d14	1790.00
gen U	8.565d14	1.155d14	1317.35
strom u	8.590d14	7.398d13	1317.92
vil U	8.690d14	1.011d14	1321.46
ans 6	9.101d14	7.242d13	1102.00
td1 4	1.094d15	1.242d14	779.49
ans 5	1.203d15	1.241d14	720.83
td1 3	1.268d15	1.777d14	687.11
td1 2	1.526d15	2.580d14	630.33

Table 8.3. (*cont.*)

Filter identifier	Effective frequency (Hz)	Frequency bandwidth (Hz)	MFD FWHM (Jy)
ans 3	1.666 d 15	1.384 d 14	598.64
td1 1	1.916 d 15	4.085 d 14	459.58
ans 2	1.935 d 15	1.866 d 14	435.19
ans 1	1.940 d 15	6.381 d 13	447.18

In this table the shorthand notation $m\,\mathrm{d}\,n$ has been used where $m\,\mathrm{d}\,n \equiv m \times 10^n$.

Sources for Table 8.3

1. IRAS (Infrared Astronomical Satellite) photometry (iras) at 12 μm, 25 μm, 60 μm and 100 μm for the star Vega (Cohen *et al.*, 1992).[51]
2. Johnson–Cousins–Glass (jcg). Monochromatic flux densities for the nine bands (U, B, V, R, I, J, H, K, L, M) are those given by Bessell (2001) for an A0V star with magnitude V = 0.00.
3. The 2 Micron All Sky Survey (2MASS) measured monochromatic flux densities in three infrared bands (J, H, K_S). Cohen *et al.* (2003) provided the absolute calibration for a zero-magnitude A0V star.
4. DENIS (den). Fouqué *et al.* (2000) used a synthetic spectrum of Vega to derive monochromatic flux densities at magnitude zero for the three bands (i, J, K_S).
5. The Vilnius (vil) seven-colour (U, P, X, Y, Z, V, S) absolute monochromatic flux densities given by Straižys (1992) in $\mathrm{W\,.\,cm^{-2}\,.\,\mu m^{-1}}$ units for an A0V star were converted to janskys using the transformation given in example 2 above.
6. Geneva (gen) seven-colour (U, B, B1, B2, V, V1, G) photometric bandwidths with their absolute calibration for Vega were determined by Rufener & Nicolet (1988), who used the relationship:

$$\log(E_\nu) = -0.4\,(m_\nu - K_\nu) + C \tag{8.34}$$

   where E_ν is the spectral irradiance at frequency ν in $\mathrm{W\,.\,m^{-2}\,.\,Hz^{-1}}$, m_ν is the measured magnitude at frequency ν, K_ν is the colour index of the Geneva band at frequency ν relative to that at the effective frequency of the B filter, and C is the zero-point shift for the B filter.
7. Strömgren (strom) photometry absolute calibrations of monochromatic irradiance are given by Sterken & Manfroid (1992) in units of $10^{-11}\,\mathrm{W\,.\,m^{-2}\,.\,nm^{-1}}$ for a star of spectral type A0V and magnitude V = 0.00. Example 2 above shows how a similar unit is converted to janskys.
8. Astronomical Netherlands Satellite (ans) photometry. Wesselius *et al.* (1980) related magnitudes to fluxes in $\mathrm{W\,.\,m^{-2}\,.\,nm^{-1}}$ (f_λ) using:

$$f_\lambda = 10^{\left(\frac{26.1+m_\lambda}{-2.5}+9\right)} \tag{8.35}$$

   which may be converted to janskys using Equation (8.22). From the ultraviolet magnitudes listed by Wesselius *et al.* (1982), the following magnitudes were extracted for Vega; ans1 = −0.491, ans2 = −0.441, ans3 = −0.462, ans4 = no measurement, ans5 = 0.046, ans6 = 0.191. Substituting these magnitudes into Equation (8.35) give values for f_λ, which may then be converted to f_ν in janskys via Equation (8.22). These monochromatic flux density values are shown in Table 8.3.
9. The Thor–Delta (td1) ultraviolet catalogue (Thompson *et al.*, 1978) the wavelength-space monochromatic flux densities of Vega in units of $10^{-10}\,\mathrm{erg\,.\,cm^{-2}\,.\,s^{-1}\,.\,Å^{-1}}$, which were converted to janskys in Table 8.3 as per example 3 above.

[51] See also http://lambda.gsfc.nasa.gov/product/iras/docs/exp.sup/ch6/C2a.html

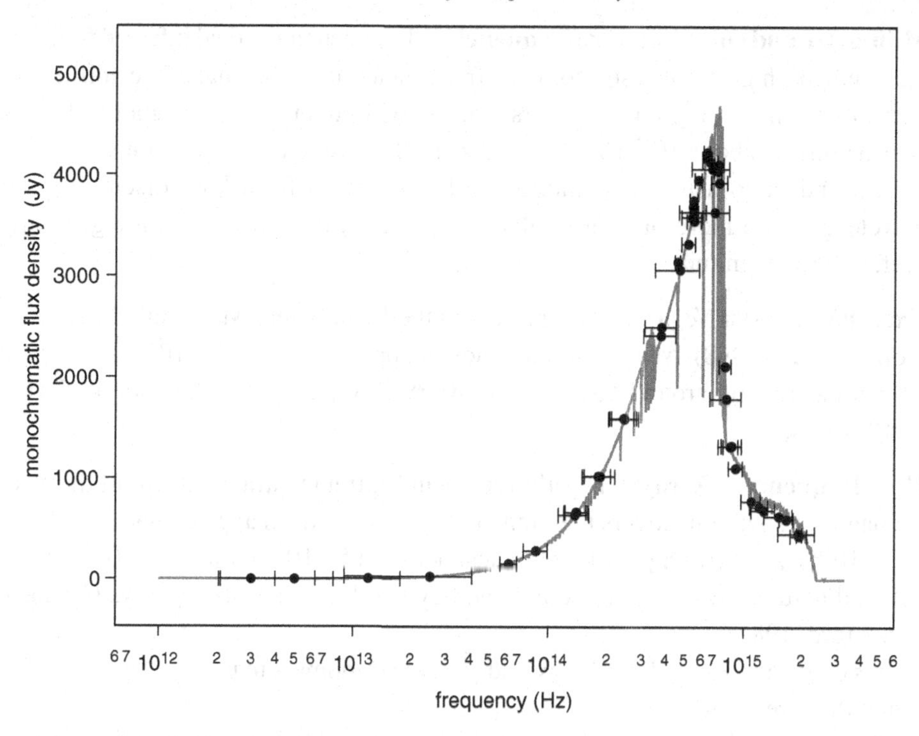

Figure 8.5. HST spectrophotometry of Vega overlaid with broad- and intermediate-band photometry. The points are at the location of the effective frequency of the filter, whose monochromatic flux density is plotted with the error bars showing the FWHM bandwidth. The low-frequency end of the spectrum in the infrared shows the location of the IRAS satellite photometric bands, the near infrared has 2MASS, DENIS and Johnson–Cousins–Glass photometry and bands, the optical part of the spectrum is overlaid with Vilnius, Geneva, Strömgren, and Johnson–Cousins–Glass photometric bands and the ultraviolet spectrum has measures and bandwidths from the TD1 and ANS satellites.

jansky, appropriately named after one of the pioneers of radio astronomy, Karl Jansky.

The monochromatic flux density per unit solid angle (Jy . rad^{-1}) is called the **brightness** of the (extended) source.

In short, radio astronomers already use derived SI units that meet with the approval of the IAU.

By way of an example, an attempt was made by Hollis *et al.* (1985) to detect Vega at radio frequencies using the VLA (Very Large Array) of the US National Radio Astronomy Observatory at a frequency of 4.86 GHz ($\lambda \sim 6$ cm) with a beamwidth of 40 μrad by 17 μrad (8.5 arcsec × 3.5 arcsec). No signal significantly greater than the background noise (30 μJy) was detected.

Millimetre and submillimetre frequencies: Between the optical infrared frequencies and the high radio astronomical frequencies is to be found the millimetre and submillimetre region. It covers a band of frequencies from about 10^{14} Hz ($\lambda \sim 3$ mm) to about 10^{15} Hz ($\lambda \sim 300\,\mu$m). This region is transitional between optical and radio astronomy and, as such, uses units from both disciplines. So wavelengths are more often used than frequencies, though janskys are generally preferred to magnitudes.

Example: the star Vega was observed with the James Clerk Maxwell telescope by Holland *et al.* (1998). At a mean frequency of approximately 3.5×10^{11} Hz, a signal strength averaging around 46 mJy was observed in a beam of diameter 48.5 μrad ($\sim 10''$ diameter).

High frequencies, X-rays: The ultraviolet and extreme ultraviolet part of the electromagnetic spectrum covers the range in frequencies from approximately 10^{15} Hz to 5×10^{16} Hz. At still higher frequencies, between 5×10^{16} Hz and 2×10^{20} Hz, the soft and hard X-rays are to be found and beyond that to 10^{26} Hz, γ-rays (Culhane & Sanford, 1981).

It is conventional for both X-ray and γ-ray astronomers to use electron volts E_ν, rather than frequency ν.

The electron volt is classed as a non-SI unit accepted for use with the international system, whose value in SI units is obtained by experiment.

The **electron volt** has the dimension of energy $[L]^2.[M].[T]^{-2}$, may be expressed in terms of base SI units as $\mathbf{m^2.kg.s^{-2}}$ and has symbol **eV**.

The relationship between electron volts and hertz is linear:

$$\nu = \frac{E_\nu}{h} = n\frac{E_0}{h} \tag{8.36}$$

where h is Planck's constant and equal to $6.626\,075\,5 \times 10^{34}$ J . s (Cox, 2000), E_0 is the energy equivalent of 1 eV in joules ($1.602\,177 \times 10^{-19}$ J) and n is the number of electron volts.

Example 1: convert 1 eV to Hz

$$\nu = 1 \times \frac{1.602\,177 \times 10^{-19}}{6.626\,076 \times 10^{-34}} = 2.418 \times 10^{14}\,\text{Hz} \tag{8.37}$$

Example 2: convert 1 keV, 1 MeV and 1 Gev to Hz

$$\nu_{(1\,\text{keV})} = 2.418 \times 10^{17}\,\text{Hz}; \quad \nu_{(1\,\text{MeV})} = 2.418 \times 10^{20}\,\text{Hz};$$
$$\nu_{(1\,\text{GeV})} = 2.418 \times 10^{23}\,\text{Hz} \tag{8.38}$$

where 1 keV $= 1000$ eV, 1 MeV $= 10^6$ eV and 1 GeV $= 10^9$ eV.

A typical observational X-ray spectrum has photon energy in keV as the abscissa plotted against the monochromatic flux density in units of photons . cm^{-2} . keV^{-1} as ordinate.

Example 3: convert photons . cm^{-2} . keV^{-1} to W . m^{-2} . Hz^{-1} and janskys

$$\begin{aligned} 1\,\text{ph}\,.\,\text{cm}^{-2}.\text{s}^{-1}.\text{keV}^{-1} &= 1\,\text{ph}\,.\,(10^{-2}\,\text{m})^{-2}.\text{s}^{-1}.(2.418\times10^{17}\,\text{Hz})^{-1} \\ &= (10^{4}\times(2.418\times10^{17})^{-1})\,\text{ph}\,.\,\text{m}^{-2}.\text{s}^{-1}.\text{Hz}^{-1} \\ &= 4.136\times10^{-14}\,\text{ph}\,.\,\text{m}^{-2}.\text{s}^{-1}.\text{Hz}^{-1} \end{aligned} \tag{8.39}$$

Now the energy of a 1 keV photon is equivalent to $1.602\,177\times10^{-16}$ J, hence:

$$\begin{aligned} 1\,\text{ph}\,.\,\text{cm}^{-2}.\text{s}^{-1}.\text{keV}^{-1} &= (1.602\,177\times10^{-16}\,\text{J})(4.136\times10^{-14})\,\text{m}^{-2}.\text{s}^{-1}.\text{Hz}^{-1} \\ &= 6.6266\times10^{-30}\,\text{J}.\text{s}^{-1}.\text{m}^{-2}.\text{Hz}^{-1} \\ &= 6.6266\times10^{-30}\,\text{W}.\text{m}^{-2}.\text{Hz}^{-1} \qquad (8.40) \\ &= 6.6266\times10^{-4}\,\text{Jy} \qquad (8.41) \end{aligned}$$

X-ray observations of Vega

An attempt to extend observational measurements of the electromagnetic spectrum of Vega to frequencies higher than the extreme ultraviolet was made by Pease *et al.* (2006) using the high-resolution camera (HRC-1) of the CHANDRA X-ray observatory. Exposure times of up to 25 900 s produced total counts of around 10 photon events over the 156-pixel array used. The energy range of the detector extended from 0.08 keV to 10.0 keV.[52] The pixel size is given as $6.4\times6.4\ \mu m^2$, so the total detector area is $6.390\times10^{-5}\ cm^{-2}$. The rate of X-ray photons incident upon the detector was $10/2599 = 3.861\times10^{-4}\ s^{-1}$ and the X-ray bandwidth taken as approximately 10 keV. So the monochromatic flux density, f_X, in ph . cm^{-2} . s^{-1} . keV^{-1} is:

$$f_X = \frac{3.861\times10^{-4}}{6.39\times10^{-5}\times10} \simeq 0.6\,\text{ph}\,.\,\text{cm}^{-2}.\text{s}^{-1}.\text{keV}^{-1} \tag{8.42}$$

which, by using the conversion in Equation (8.41), is equivalent to 4.0×10^{-4} Jy. The centre of the measured X-ray band was taken to be 5 keV or 1.209×10^{18} Hz.

High frequencies: γ-rays are photons with the highest detectable energies. Ground-based instruments, such as the HESS (High Energy Stereoscopic System) array in Namibia,[53] were designed to detect 1 TeV ($10^{12}\text{eV} \equiv 2.416\times10^{26}$ Hz)

[52] See http://cxc.harvard.edu/proposer/POG/html/chap7.html page 2 of 19
[53] See http://www.saao.ac.za/~wgssa/as5/steenkamp.html

photons with a monochromatic flux density as low as 10^{-12} erg . s^{-1} . cm^{-2} ($\equiv 10^{-15}$ W . m^{-2}) indirectly by means of the Čerenkov radiation emitted by the γ-ray's passage through the Earth's atmosphere. The use of several telescopes in the array allows the direction from which the γ-ray photon originated to be estimated.

For lower-energy γ-ray photons, detectors are commonly used with high-altitude balloons, rockets and satellites. The most recent project is the Fermi γ-ray space telescope mission launched by NASA in 2008. The principal instrument on this satellite is the Large Area Telescope (Atwood *et al.*, 2009), which covers the energy range from 20 MeV to 300 GeV, corresponding to frequencies from 4.8×10^{21} Hz to 7.5×10^{25} Hz.

In the first catalogue of active galactic nuclei detected by the Fermi Large Area Telescope, Abdo *et al.* (2010), in common with most γ-ray astronomers, used photons . cm^{-2} . s^{-1} as a unit for flux density, with the electron volt being used instead of frequency in hertz. Atwood *et al.* (2009) used a mixture of units for flux density, such as photons . cm^{-2} . s^{-1} and particles . m^{-1} . s^{-1}.

The unit transformations used in the section on X-ray astronomy apply equally well to the γ-ray units.

Extended flux-density plot for Vega

The extra values, or upper limits, for monochromatic flux densities found for radio, millimetre, submillimetre, X-rays and γ-rays have been added to the data used to construct Figure 8.5 to produce Figure 8.6. To accommodate the larger range of monochromatic flux densities in the enhanced data set, a logarithmic scale is used for the ordinate.

Wavenumbers of electromagnetic radiation may be defined as the number of electromagnetic waves per unit length.

The SI unit for **wavenumber** is the **inverse metre** with dimension $[L]^{-1}$ and symbol $\mathbf{m^{-1}}$.

In infrared astronomy a cgs wavenumber unit commonly found is the **kayser** or inverse centimetre (cm^{-1}). Occasionally, the inverse micrometre, μm, is used. The IAU recommends only the use of the m^{-1}. The relationship between frequency ν, wavelength λ, and wavenumber $\overline{\nu}$, is given by Browning (1969) as:

$$\overline{\nu} = \frac{1}{\lambda} = \frac{\nu}{c} \tag{8.43}$$

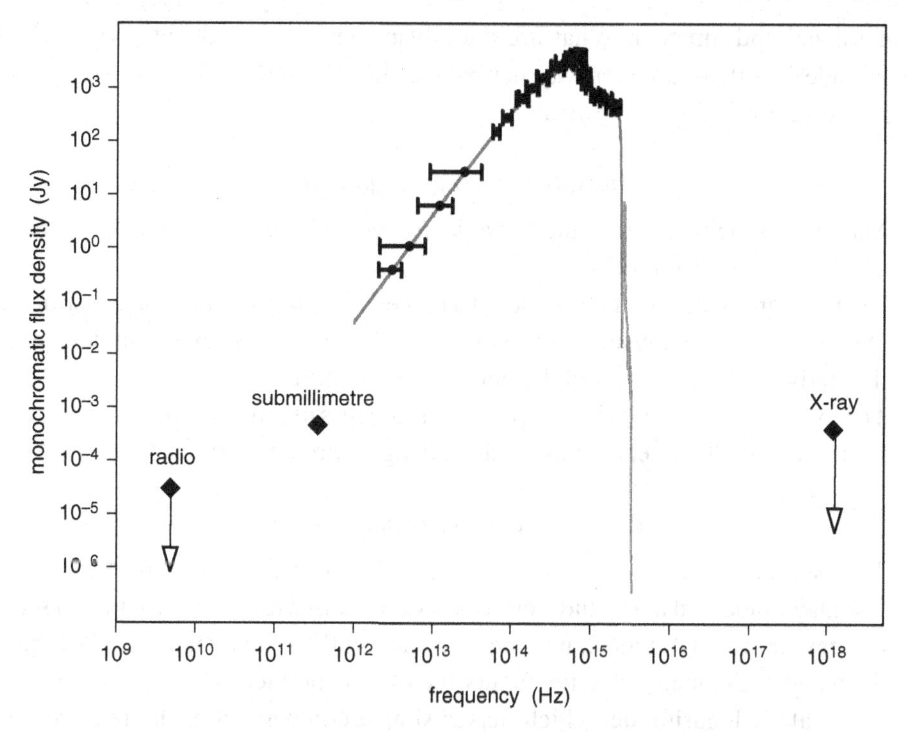

Figure 8.6. A log-log plot of HST spectrophotometry of Vega overlaid with broad- and intermediate-band ultraviolet, optical and infrared photometry, with diamond symbols for measures or attempted measures in the radio, submillimetre and X-ray bands. The radio and X-ray plots are upper limits to the flux at that frequency.

Some sample conversions: Consider the rest wavelength, $\lambda_\alpha = 6562.817$ Å (Moore, 1959) of the hydrogen H_α line of the Balmer series:

$$6\,562.817\,\text{Å} = 6.562\,817 \times 10^{-7}\,\text{m}$$

$$= 6.562\,817 \times 10^{-5}\,\text{cm}$$

$$\overline{\nu}_{\text{m}} = \frac{1}{6.562\,817 \times 10^{-7}}\,\text{m}^{-1} = 1.523\,736 \times 10^{6}\,\text{m}^{-1} \tag{8.44}$$

$$\overline{\nu}_{\text{cm}} = \frac{1}{6.562\,817 \times 10^{-5}}\,\text{cm}^{-1} = 15\,237\,\text{cm}^{-1} \tag{8.45}$$

8.3 Magnitudes

A very brief history (for more detail, see Hearnshaw 1996) of the use of magnitudes in describing the brightness of stars is given in Chapter 12 and the present chapter includes numerous examples of different magnitude schemes used over the past century. Magnitudes are thoroughly engrained in the minds of astronomers, both professional and amateur. What are the advantages and disadvantages of using magnitudes? Is there a better way that would link magnitudes more strongly and logically to physical measurements?

The advantages of magnitudes

1. Magnitudes, being logarithmic, are able to cover, with a small range of numbers, a large range of brightness.
2. Using magnitudes and magnitude differences, simple plots may be constructed that often provide considerable astrophysical insight to a given problem, e.g., the study of star clusters and the interstellar medium.
3. The overwhelming majority of optical astronomers use magnitudes, as they have for more than 2000 years, making any change extremely difficult.

The disadvantages of magnitudes

1. The magnitude scale was originally conceived as an ordinal scale (1st magnitude, 2nd magnitude, 3rd magnitude, etc.) that was later transformed to a cardinal scale (magnitude 1, magnitude 2, magnitude 3, etc.) without inversion (i.e., the bigger the value of the magnitude the fainter the object and vice versa).
2. The scale is logarithmic, which makes simple comparison of the relative flux densities of two or more astronomical objects difficult.
3. The logarithmic base was selected as the $\sqrt[5]{100}$ or approximately 2.511 886 to tie the ordinal and cardinal schemes together. The difference in brightness between a star of the first magnitude and one of the sixth magnitude was set equal to a factor of 100, leading to a one magnitude difference equalling $\sqrt[5]{100}$.
4. The zero point of a magnitude scale is chosen arbitrarily.

Given the uses to which magnitudes are put and their obvious universality it would seem appropriate to attempt to bridge the gap between radiometry and astronomical magnitudes.

8.3.1 A proposed magnitude system based on radiometric measures

In proposing a new magnitude scale, attention needs to be paid to the lists of advantages and disadvantages given above. The current magnitude scale is defined mathematically as:

$$m = -2.5\log(f) + m_0 \tag{8.46}$$

where m is the magnitude and f is a number related to the flux density of the object being measured. This number may be counts obtained from a photoelectric photometer or CCD camera, or a measure obtained from a photographic plate (e.g., image area), or even a comparative visual estimate (see Chapter 12). m_0 is the

Table 8.4. *Magnitude differences for various flux-density differences* $\left(\frac{f}{f_0}\right)$ *with varying logarithmic bases*

		Δm		
$\frac{f}{f_0}$	$\log_2$	$\log_{10}$	$\log_{\sqrt[5]{100}}$	ln
2	1.0000	0.3010	0.7500	0.6931
10	3.3219	1.0000	2.5000	2.3026
50	5.6438	1.6990	4.2474	3.9120
100	6.6439	2.0000	5.0000	4.6052
10000	13.2877	4.0000	10.0000	9.2103

zero-point adjustment that transforms the measured magnitude of the detector into that of a standard magnitude system.

Essentially, the simple revised magnitude system proposed changes Equation (8.46) to:

$$m = k\,(\text{logarithm})_{\text{base}}\,(f_{\text{janskys}}) \tag{8.47}$$

The base of the logarithm has to be selected, plus a constant coefficient, k. Four possibilities for the logarithm base are, 2, 10, $\sqrt[5]{100}$, and e, where e is defined by Abramowitz & Stegun (1972) to be:

$$\mathrm{e} = \lim_{n\to\infty}\left(1+\frac{1}{n}\right)^n = 2.718\,281\,828\,4... \tag{8.48}$$

The magnitude changes in each system corresponding to a measured flux density change of 2×, 10×, 50×,100× and 10 000× are shown in Table 8.4.

There is an obvious advantage in staying with the $\sqrt[5]{100}$ system in that the values of differential magnitude will remain as they are now, $\log_{10}$ is the most commonly used system of logarithms, $\log_2$ is of particular significance in computing and $\ln = \log_e$ is of significance in many branches of mathematics and physics.

Naturally, any decision on what form, if any, a new magnitude system should take is for the IAU to decide. In order to provide meaningful worked examples on how such a revised system might operate, the logarithmic base e has been selected. An appropriate name might be **e-magnitudes**, ***em***, which would be defined as:

$$em_\nu = \ln(f_\nu) \tag{8.49}$$

where ν is the effective frequency of the electromagnetic radiation being measured and f_ν is the monochromatic flux density at frequency ν measured in janskys.

Example 1: e-magnitude computation.
Derive the e-magnitude for an A0V star with $m_V = 0.00$ and $f_\nu = 3636$ Jy (Bessell, 2001).

$$em_V = \ln(f_\nu) = \ln(3636) = +8.199 \tag{8.50}$$

Apparent and absolute magnitudes

The observed, or **apparent magnitude**, or monochromatic flux density, depends on the intrinsic brightness of the celestial object, its distance and the amount of interstellar absorption along the line of sight to the object. Interstellar extinction and reddening is caused by the presence of dust grains that absorb and scatter the radiation from the celestial object. **Absolute magnitude** is defined to be the magnitude of an unreddened celestial body at a distance of 10 pc ($3.085\,677\,6 \times 10^{17}$ m).

The **colour index** is defined to be the difference between two magnitudes, determined from observing a celestial object through two different bandwidths, e.g., (B–V), (R–I). In radiometric terms, the colour index is the ratio of the observed monochromatic flux densities in two frequency bands and is thus dimensionless, e.g.,

$$(\text{colour index})_{f_\nu} = \frac{f_{\nu_B}\,(\text{Jy})}{f_{\nu_V}\,(\text{Jy})} \tag{8.51}$$

A list of absolute magnitudes for stars of spectral type (see Chapter 12) O5 to M5 in the Johnson–Cousins–Glass broadband photometry is given in Cox (2000). These magnitudes were converted to janskys, as shown in Example 2 below, using the material in Table 8.3 for jcgU, jcgB and jcgV. Subsequently, these monochromatic flux densities were transformed into e-magnitudes and e-colours.

Example 2: absolute e-magnitude and e-colour computations:
convert the UBV magnitudes for a G0V star to e-magnitudes and e-colours.

For a G0V star, the following absolute magnitudes may be taken or computed from the values listed in Cox (2000): U = +5.04; B = +4.98; V = +4.40. Convert these magnitudes into janskys (f_U, f_B and f_V) using the monochromatic flux densities given by Bessell (2001) for an A0V star with zero magnitude and zero colour indices (Equation 8.51).

$$f_{U_0} = 1790\,\text{Jy} \quad f_{B_0} = 4063\,\text{Jy} \quad f_{V_0} = 3636\,\text{Jy} \tag{8.52}$$

$$f_U = f_{U_0} \times 10^{-0.4U} \tag{8.53}$$

$$f_B = f_{B_0} \times 10^{-0.4B} \tag{8.54}$$

$$f_V = f_{V_0} \times 10^{-0.4V} \tag{8.55}$$

For the G0V star, the computed fluxes are:

$$f_U = 17.25\,\mathrm{Jy} \qquad f_B = 41.39\,\mathrm{Jy} \qquad f_V = 63.19\,\mathrm{Jy} \tag{8.56}$$

The e-magnitudes corresponding to these fluxes are:

$$em_U = \ln(17.25) = +2.85$$
$$em_B = \ln(41.39) = +3.72$$
$$em_V = \ln(63.19) = +4.15 \tag{8.57}$$

Note that in the e-magnitude system the **bigger** the numerical value of the e-magnitude the **brighter** the astronomical object.
e-colours derived from the e-magnitudes are:

$$e(B-V) = 3.72 - 4.15 = -0.43$$
$$e(U-B) = 2.85 - 3.72 = -0.87 \tag{8.58}$$

Plots of the original data from Cox (2000) and Straižys (1992) in the form of colour–magnitude and colour–colour diagrams are shown in Figures 8.7 and 8.8, whilst similar plots derived from monochromatic flux densities are shown as e-colour–magnitude and e-colour–colour diagrams in Figures 8.9 and 8.10.

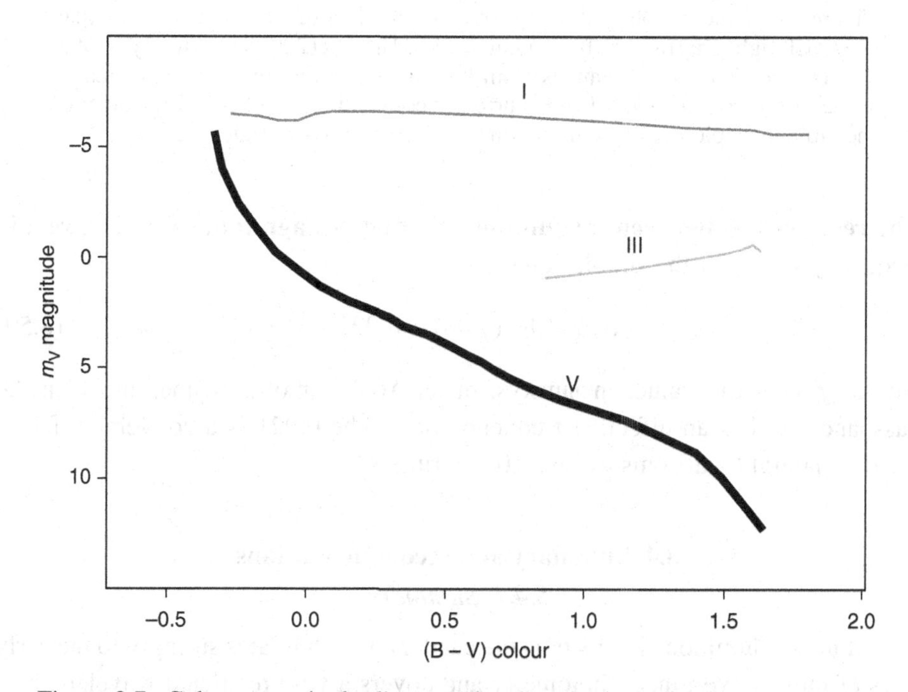

Figure 8.7. Colour–magnitude diagram showing the loci of luminosity class I (dark grey), III (light grey) and V (black) stars as listed by Cox (2000) and Straižys (1992).

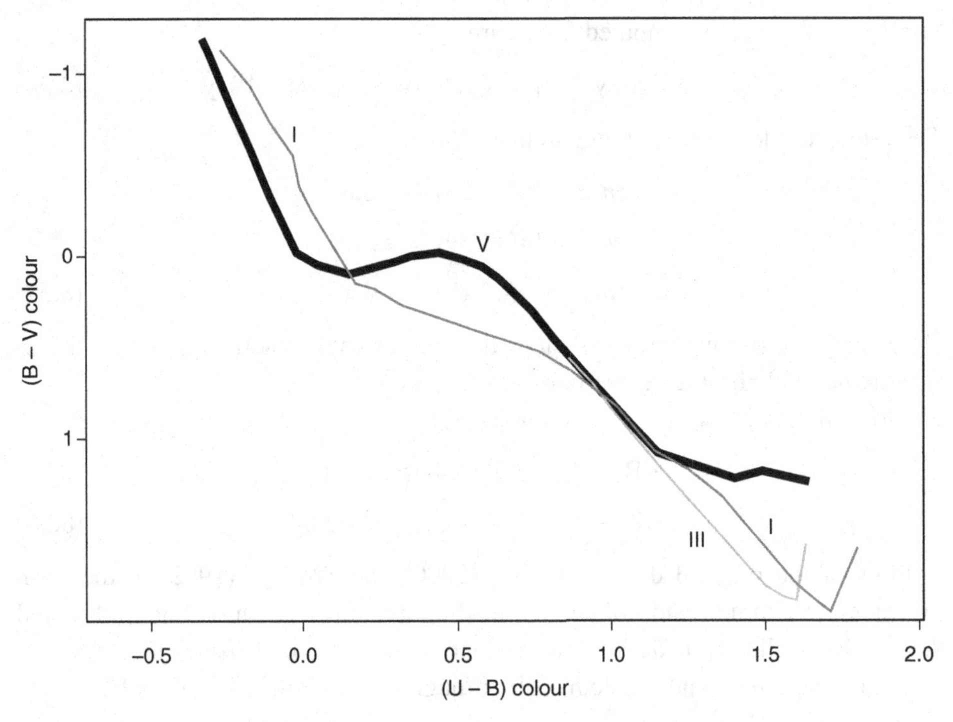

Figure 8.8. Colour–colour diagram showing the loci of luminosity class I (dark grey), III (light grey) and V (black) stars as listed in Cox (2000) and Straižys (1992). The colour of the stars changes from blue to red in the direction of increasing value of the colour index. In the currently conventional colour–colour diagram, the ordinate is plotted in the direction of decreasing (B–V) index.

The relationship between magnitudes m_ν, and e-magnitudes em_ν is given, in frequency space, by the simple equation:

$$em_\nu = \ln((f_\nu)_0) - 0.921\, m_\nu \tag{8.59}$$

where $(f_\nu)_0$ is the value, in janskys, of an A0V star of zero magnitude in the passband that has an effective frequency of ν. The 0.921 is a conversion factor relating natural logarithms to base-10 logarithms.

8.4 Summary and recommendations

8.4.1 Summary

The SI unit of luminous intensity is the candela, which relates strongly to the early days of human-eye-tuned photometry and covers a very restricted wavelength or frequency range. In parallel with the pure photometric system of units are those of radiometry and astronomy. The radiometric units are directly related to the mass,

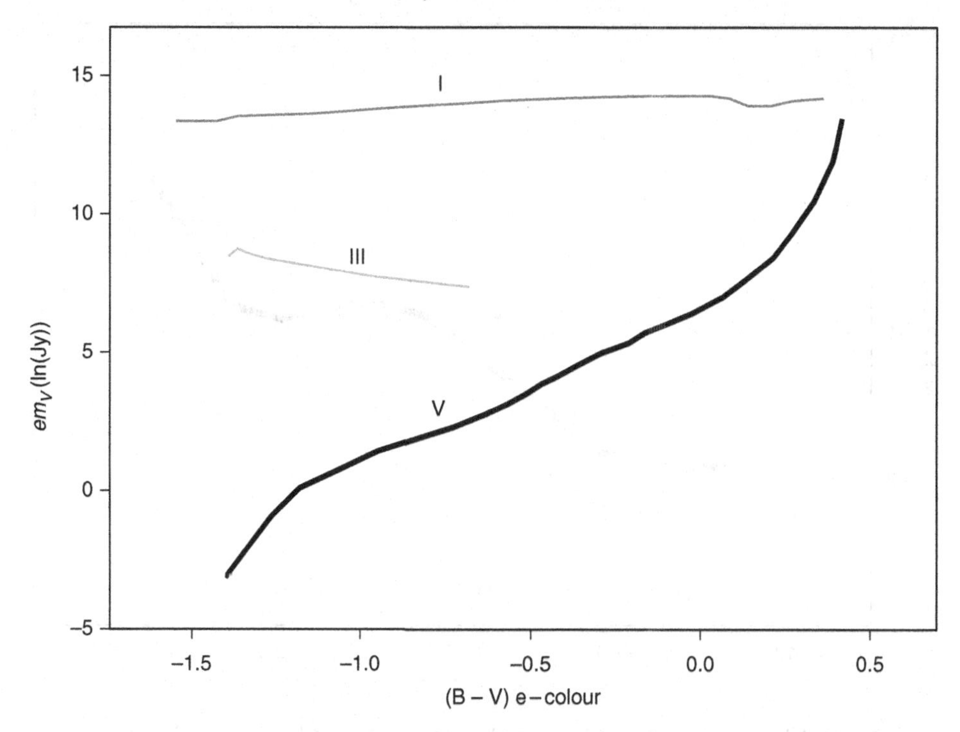

Figure 8.9. e-colour–magnitude diagram showing the loci of luminosity class I (dark grey), III (light grey) and V (black) stars derived using Equations (8.51) to (8.57). Stars increase in brightness with increasing value of e-magnitude and their colour changes from red to blue with increasing value of e-colour.

length and time base units of the SI system. Astronomical units are rather more esoteric. Comparisons between the different photometries are given and worked examples of converting from one set of units to another given. For many of the common astronomical photometric systems, the central wavelengths and bandwidths are transformed from wavelength to frequency space. Examples are given of determining the spectral irradiance (radiometic unit) and the monochromatic flux density of stars from many different published non-SI unit sources. Magnitudes, a much-used astronomical photometric unit, are converted to monochromatic flux densities and a flux versus frequency plot of HST spectrographic observations of Vega produced.

Some selected commonly used photometric systems from the ultraviolet to the infrared are calibrated in radiometric terms and in janskys. Very high frequency astronomical observations in the X-ray and γ-ray regions are also converted to janskys, with observations from the opposite end of the electromagnetic spectrum in the submillimetre, millimetre and radio regions. Wavenumbers of three different

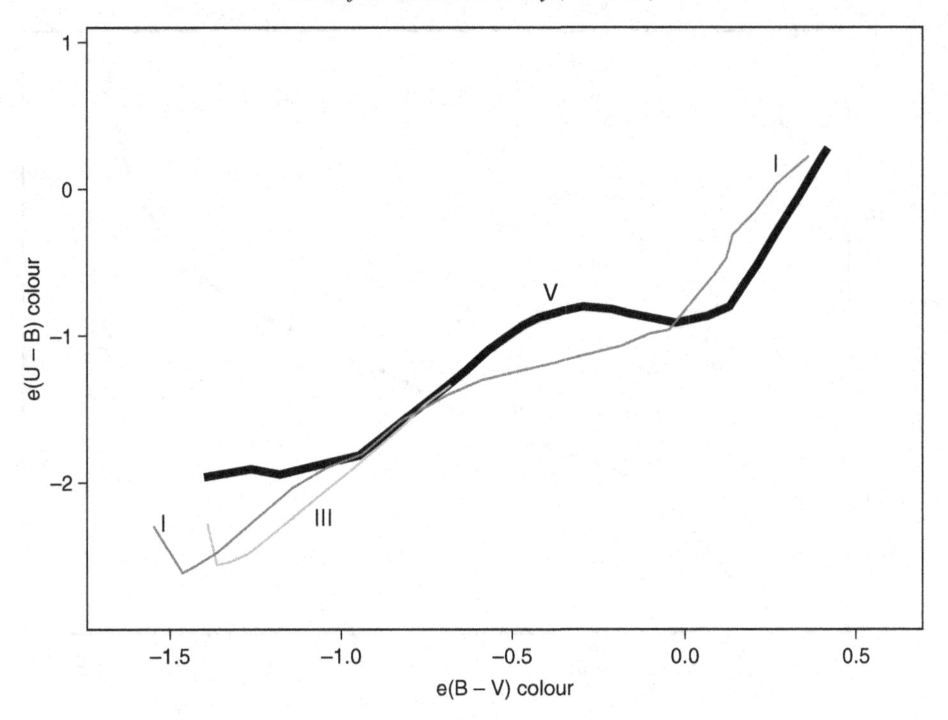

Figure 8.10. e-colour–colour diagram showing the loci of luminosity class I (dark grey), III (light grey) and V (black) stars as listed by Cox (2000). The stars colour change from red to blue in the direction of increasing colour index along both axes.

types, μm^{-1}, cm^{-1} and m^{-1} are converted to frequencies in hertz and simple worked examples given.

A discussion on the advantages and disadvantages of astronomical magnitudes is given along with a proposed magnitude system based on radiometric measures in janskys.

8.4.2 Recommendations

The major shortcoming of the defined SI photometric unit, the candela, when applied to astronomy, is due principally to the restricted frequency band covered. However, the radiometric units defined in terms of base SI units are easily related to astronomical photometry. The great variety of different units in common use in astronomical photometry makes inter-comparison between them time consuming and prone to the making of mistakes.

For observational work, the most logical unit to use for monochromatic flux densities is the jansky. It has been in use for many decades by both radio and infrared astronomers and is not difficult to calculate from other similar units, which may differ by scale factors or by being tied to wavelength rather than frequency space. The logarithm (whether natural or to base 10) of the flux density in janskys makes a very acceptable physically based magnitude that allows for the continued use of colour–magnitude and colour–colour diagrams. For both theoretical and observational studies of the total power output of all types of astronomical bodies, the watt is ideal.

9
Unit of thermodynamic temperature (kelvin)

9.1 SI definition of the kelvin

The **kelvin**, unit of thermodynamic temperature, is the fraction 1/273.16 of the thermodynamic temperature of the triple point of water.

The dimension of thermodynamic temperature is [**Θ**], its unit is the **kelvin** and its symbol is **K**.

9.1.1 Possible future definition of the kelvin

Under discussion at the present time is the redefining of the unit of temperature in the following way:

The kelvin is a unit of thermodynamic temperature such that the Boltzmann constant is exactly $1.380\,650\,5 \times 10^{-23}\,\mathrm{J\,.\,K^{-1}}$ (joules per kelvin).

9.1.2 Definition of temperature

The average kinetic energy of the molecules, atoms and ions that comprise an object may be used as a measure of its thermodynamic temperature.

Given two objects with different temperatures, the one with the higher temperature (the hotter object) will transfer heat to the object with the lower temperature (the cooler object) by means of radiation, convection or conduction, depending on the nature of the objects and their location, until the temperatures of the objects are equalized.

9.1.3 Thermodynamic temperature

The thermodynamic temperature, T_t, of a system may be defined as the inverse of the rate of change of entropy S, with respect to the internal energy U, assuming that the volume V of the system and the number of its constituent parts N remain constant,[54] i.e.

$$T_t = \left(\frac{\partial S}{\partial U}\right)^{-1}_{N,V} \tag{9.1}$$

The entropy S, is a measure of the disorder of the system and increases with increasing levels of disorder (e.g., steam has a higher entropy than water, which in turn has a higher entropy than ice).

9.1.4 Kinetic temperature

The kinetic temperature, T_k, may be defined for an ensemble of identical particles of individual mass m, with root-mean-square particle speed $< v^2 >$ as:

$$T_k = \frac{m < v^2 >}{3k} \tag{9.2}$$

where the speed distribution $f(v)$, of the particles is Maxwellian,[55] i.e.,

$$f(v) = 4\pi \left(\frac{m}{2\pi\, k\, T}\right)^{\frac{3}{2}} v^2 \mathrm{e}^{-\frac{m v^2}{2kT}} \tag{9.3}$$

Example: check the dimensional consistency in Equation (9.2)
The dimension of the LHS of Equation (9.2) is $[\Theta]$. The RHS of Equation (9.2) involves Boltzmann's constant k, which has $\mathrm{J\,.\,K^{-1}}$ as a unit and may be expressed in dimensional terms as $[M]\,.\,[L]^2\,.\,[T]^{-2}\,.\,[\Theta]^{-1}$, hence:

$$\begin{aligned}\dim[\mathrm{RHS}] &= \frac{[M]\,.\,[L]^2\,.\,[T]^{-2}}{[M]\,.\,[L]^2\,.\,[T]^{-2}\,.\,[\Theta]^{-1}}\\ &= [\Theta]\\ &= \dim[\mathrm{LHS}]\end{aligned}$$

9.2 Temperature scales

Various temperature scales have been used over the past two hundred years, of which the Celsius, Fahrenheit, Kelvin and Rankine scales are still in common use.

The differences between such scales are essentially due to different temperature unit size and different zero points.

54 See http://hyperphysics.phy-astr.gsu.edu/hbase/thermo/temper2.html
55 See http://hyperphysics.phy-astr.gsu.edu/hbase/kinetic/kintem.html

9.2.1 Converting the Celsius, Fahrenheit and Rankine scales to the Kelvin temperature scale

Temperatures measured in the Celsius and Fahrenheit scales may be either positive or negative and both define the freezing point of water with a pre-assigned value for temperature (0 for celsius and 32 for fahrenheit). The kelvin and rankine have as their lowest temperature value 0, known as **absolute zero**, the temperature at which, in classical mechanical systems, all translational motion of atoms and molecules ceases. In quantum mechanical systems there remains the possibility of zero-point energy-induced particle motion.

The temperature intervals in the Kelvin and Celsius scales are the same, 1 K = 1 °C, as are those in the Fahrenheit and Rankine scales, 1 °R = 1 °F. The Kelvin and Rankine scales are both absolute temperature systems.

The IAU recommends citing temperatures in kelvin. To assist with their recommendation, the appropriate conversion formulae from one temperature system to another is given. There are many online converters such as that given in the *Engineering Toolbox*.[56]

To convert Celsius temperatures to Kelvin temperatures

$$\theta_K = \theta_C + 273.16 \tag{9.4}$$

where θ_K is the temperature in kelvin and θ_C the temperature in degrees Celsius.

To convert Fahrenheit temperatures to Kelvin temperatures

$$\theta_K = \frac{5\,(\theta_F - 32)}{9} + 273.16 \tag{9.5}$$

where θ_F is the temperature in degrees Fahrenheit.

To convert Rankine temperatures to Kelvin temperatures

$$\theta_K = \frac{5\,\theta_R}{9} \tag{9.6}$$

where θ_R is the temperature in degrees Rankine.

In astronomy K, °C and °F are all commonly used. Fahrenheit is used mainly in popular astronomical publications in countries where that scale is more familiar to the local readers.

[56] See http://www.engineeringtoolbox.com/temperature-d_291.html

9.3 Some examples of the temperatures of astronomical objects

Solar system temperatures

For the Sun, planets, some dwarf planets, some satellites and Halley's comet, temperatures have been determined for their surfaces (S) or within their atmospheres (A) at a point where the atmospheric pressure is equal to 10^5 Pa. Temperatures are given in Tables 9.1 and 9.2 in the four systems, kelvin (K), celsius (C), rankine (R), and fahrenheit (F), defined above in Equations (9.4) to (9.6).

The kelvin scale values were obtained from Cox (2000) for the Sun, planets and satellites, from Emerich *et al.* (1988) for Halley's comet and websites for Ceres[57] and Eris.[58]

More extreme astronomical temperatures

The temperatures in Table 9.2 range from absolute zero through the temperature of the cosmic microwave background (CMB) radiation (the red-shifted remnant of the Big Bang event), interstellar molecular gas clouds, cool brown dwarfs, hot stellar surfaces, the central temperature of a pre-eruptive supernova star to the Planck temperature that existed a minute fraction of a second after the Big Bang. As with Table 9.1, temperatures are given in the Kelvin, Celsius, Rankine and Fahrenheit systems.

Determination of effective temperature

The effective temperature, T_{eff}, of a celestial object is defined by the equation:

$$T_{\text{eff}} = \left(\frac{L}{4\pi\,\sigma\,R^2}\right)^{\frac{1}{4}} \tag{9.7}$$

where L is the luminosity of the object in watts, R is the radius of the object in metres and σ is the Stefan–Boltzmann constant equal to $5.670\,51 \times 10^{-8}\,\text{W}\,.\,\text{m}^{-2}\,\text{K}^{-4}$. It is not uncommon in astronomy to find the Stefan–Boltzmann constant given in cgs units, with the radius and luminosity of the celestial object likewise in cgs units.

Conversion of the Stefan–Boltzmann constant from cgs to SI units

In cgs units the value of the Stefan-Boltzmann constant σ, is:

$$\begin{aligned}\sigma &= 5.67051 \times 10^{-5}\,\text{erg}\,.\,\text{s}^{-1}\,.\,\text{cm}^{-2}\,.\,\text{K}^{-4} \\ &= 5.67051 \times 10^{-5}\,(10^{-7}\,\text{W})\,.\,(10^{-2}\text{m})^{-2}\,.\,\text{K}^{-4}\end{aligned} \tag{9.8}$$

[57] See http://www.princeton.edu/~willman/planetary_systems/Sol/Ceres/
[58] See http://starchild.gsfc.nasa.gov/docs/StarChild/solar_system_level2/eris.html

Table 9.1. *Solar System temperatures*

Object	Temperature K	°C	°R	°F
Solar centre	1.6 d 7	1.6 d 7	2.88 d 7	2.88 d 7
Sun T_{eff}	5779	5506	10 400	9943
Planets				
Mercury (S)	440	167	792	332
Venus (S)	730	457	1314	854
Earth (S)	290	17	522	62
Mars (S)	226	−47	407	−53
Jupiter (A)	165	−108	297	−163
Saturn (A)	134	−139	241	−218
Uranus (A)	76	−197	137	−323
Neptune (A)	73	−200	131	−328
Dwarf planets				
Ceres (S)	173	−100	311	−148
Pluto (S)	58	−215	104	−355
Eris (S)	30	−243	54	−406
Satellites				
Moon (S)	250	−23	450	−10
Titan (S)	94	−179	169	−290
Triton (S)	38	−235	68	−391
Comet				
Halley	375	102	675	215

In this table the shorthand notation $m \, \mathrm{d} \, n$ has been used where $m \, \mathrm{d} \, n \equiv m \times 10^n$.

$$= 5.67051 \times 10^{-5} . 10^{-3} \, \mathrm{W.m^{-2}.K^{-4}}$$

$$= 5.67051 \times 10^{-8} \, \mathrm{W.m^{-2}.K^{-4}} \qquad (9.9)$$

where Equation (9.9) gives the value of the Stefan–Boltzmann constant in SI units.

Example: determine the effective temperature of the Sun

Given the luminosity of the Sun, $L_\odot = 3.845 \times 10^{26}$ W and the solar radius to be $R_\odot = 6.955\,08 \times 10^8$ m, then its effective temperature is:

$$T_{\mathrm{eff}} = \left(\frac{L_\odot}{4\pi\sigma R_\odot^2}\right)^{\frac{1}{4}}$$

$$= \left(\frac{3.845 \times 10^{26}}{4\pi \times (5.67051 \times 10^{-8}) \times (6.95508 \times 10^8)^2}\right)^{\frac{1}{4}}$$

$$= 5779 \, \mathrm{K} \qquad (9.10)$$

Table 9.2. *More extreme astronomical temperatures*

Object	Temperature K	°C	°R	°F
Absolute zero	0	−273.2	0	−459.7
Cosmic microwave background	2.7	−270.4	4.9	−454.8
Interstellar molecular cloud	10	−263	18	−442
Cool brown dwarf	450	177	810	350
T_{eff} (O5V star)	42 000	41 730	75 600	75 140
25 $M_{\odot}$ supernova	3.4 d 9	3.4 d 9	6.1 d 9	6.1 d 9
Planck temperature	1 d 32	1 d 32	1.8 d 32	1.8 d 32

In this table the shorthand notation $m\,\mathrm{d}\,n$ has been used where $m\,\mathrm{d}\,n \equiv m \times 10^n$.

Notes:

Cosmic microwave background radiation was discovered by Penzias & Wilson (1965). The value given in Table 9.2 is that obtained from measurements made using the COBE (COsmic Background Explorer) satellite by Mather *et al.* (1994).

Ferrière (2001) lists various parameters of the different components of interstellar gas for the region of space near the Sun. Temperatures are given for the molecular (the value in the table), cold and warm atomic, and warm and hot ionized gas.

Lucas *et al.* (2010) reports on the discovery of a cool brown dwarf listed as UGPS J0722-05 in the UKIDSS Galactic Plane Survey. A provisional spectral type (see Chapter 12) T10 has been assigned, though the possibility of a new spectral class has been suggested. The temperature in Table 9.2 is the mean value of the range given.

The effective temperature (see below) of the brightest main sequence star (spectral type O5V) is given as 42 000 K in Cox (2000).

Weaver *et al.* (1978) give the central temperature of a 25 $M_{\odot}$ ($= 4.97 \times 10^{31}$ kg) supernova at the onset of core ignition for the silicon-to-iron-burning stage.

The Planck temperature is that which existed at the Planck time (5.38×10^{-44}s) after the Big Bang (Lang, 2006).

9.4 Blackbody radiation

A **blackbody** may be defined as an idealized object that neither reflects nor scatters electromagnetic radiation incident upon it, but absorbs and re-emits it completely. A blackbody emits radiation as a continuous spectrum that depends solely upon its temperature and not upon its constituent parts, shape or internal structure.

Planck's law

The frequency distribution of the radiation emitted by a blackbody following Planck's law is a function of temperature alone. The spectral radiance or specific intensity, $B_\nu(T)$, emitted by a blackbody whose temperature is T K, at a frequency of ν Hz, is given by:

$$B_\nu(T) = \frac{2h\nu^3}{c^2}\left(\frac{1}{e^{\frac{h\nu}{kT}} - 1}\right) \mathrm{W\,.\,m^{-2}\,.\,Hz^{-1}\,.\,sr^{-1}} \tag{9.11}$$

where h is Planck's constant and equal to $6.626\,075 \times 10^{-34}$ J . s, k is Boltzmann's constant and equal to $1.380\,658 \times 10^{-23}$ J . K^{-1}, and c is the speed of light equal to $2.997\,924 \times 10^{8}$ m . s^{-1}.

In terms of wavelength λ, in metres, the expression for the spectral radiance or specific intensity, $B_\lambda(T)$, emitted by a blackbody whose temperature is T K, is given by:

$$B_\lambda = \frac{2hc^2}{\lambda^5}\left(\frac{1}{e^{\frac{hc}{\lambda kT}}} - 1\right) \mathrm{W.m^{-2}.m^{-1}.sr^{-1}} \tag{9.12}$$

Examples of log–log blackbody curves in frequency space for various astronomical objects are given in Figure 9.1 from the very low temperature ($T = 2.73$ K) of the cosmic background radiation to the very high temperature ($T = 3.4 \times 10^9$ K) near the centre of a 25 $M_\odot$ star.

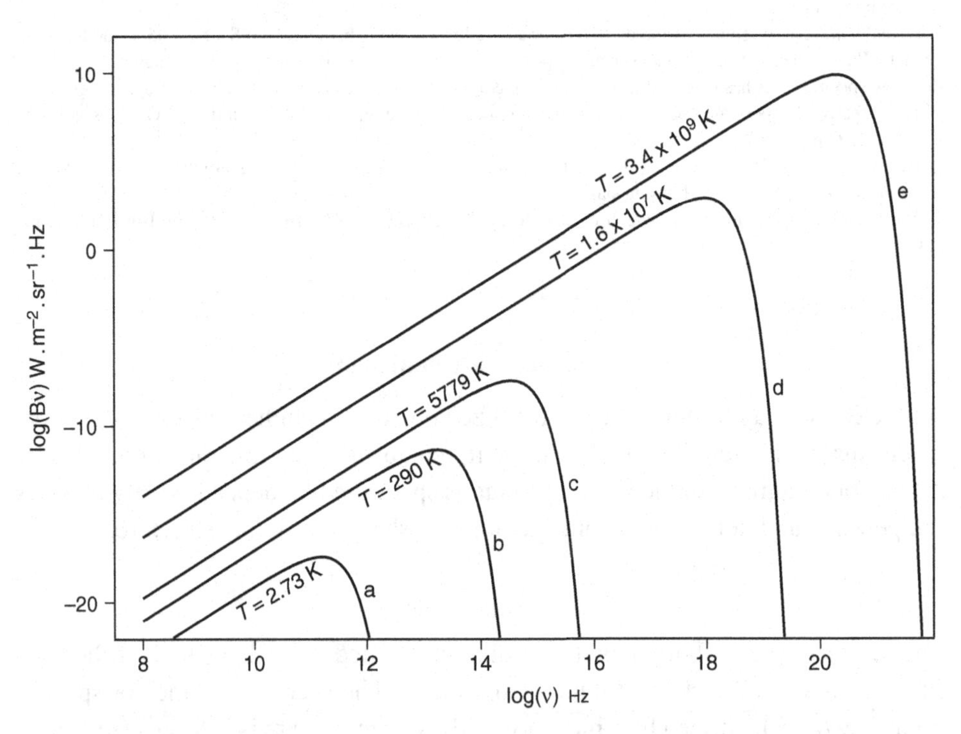

Figure 9.1. Blackbody curves computed using Equation (9.11). Curve 'a' is for a blackbody at the same temperature as the cosmic background radiation, curve 'b' for the surface of the Earth, curve 'c' for the surface of the Sun, curve 'd' for the centre of the Sun and curve 'e' for the centre of a 25 $M_\odot$ star.

9.4.1 Wein's displacement law

Wein's displacement law gives the wavelength λ_{max}, or frequency ν_{max}, of the maximum value of the specific intensity of a Planck curve. The wavelength or frequency is a function solely of the temperature, but is different in the sense that $\lambda_{max} \neq c / \nu_{max}$ due to the different forms of the Planck function in wavelength and frequency space. Values for λ_{max} and ν_{max} may be found by differentiating the appropriate expression for specific intensity with respect to either λ or ν and setting the result equal to zero.

For λ_{max} the expression is:

$$\lambda_{max} = \frac{b}{T} \tag{9.13}$$

where T is the blackbody temperature in kelvin and the Wein displacement constant $b = 0.002\,897\,8$ K . m.

For ν_{max} the expression is:

$$\nu_{max} = 5.88 \times 10^{10}\, T \tag{9.14}$$

Example: compute λ_{max} and ν_{max} for a blackbody of temperature 2.726 K

$$\lambda_{max} = \frac{0.002\,897\,8}{2.726} = 1.063 \times 10^{-3}\,\text{m} = 1.063\,\text{mm} \tag{9.15}$$

$$\nu_{max} = 5.88 \times 10^{10} \times 2.726 = 1.603 \times 10^{11}\,\text{Hz} \tag{9.16}$$

9.4.2 Stefan–Boltzmann law

The Stefan–Boltzmann law states that the power P, radiated per unit surface area of a blackbody, is directly proportional to the fourth power of its absolute temperature, so that:

$$P = \sigma T^4 \tag{9.17}$$

In SI units, P is measured in $\mathbf{W\,.\,m^{-2}}$ and T in **K**. The total power emitted by a blackbody of temperature, T, in all directions is simply:

$$P_{total} = 4\pi\sigma T^4 \tag{9.18}$$

The effective temperature, $T_\star$, of a star may be written as:

$$T_\star = \left(\frac{L_\star}{4\pi\sigma R_\star^2}\right)^{\frac{1}{4}} \tag{9.19}$$

or, by rearranging, the luminosity $L_\star$ of the star becomes:

$$L_\star = 4\pi\sigma R_\star^2 T_\star^4 \tag{9.20}$$

9.4.3 Aerial or antenna temperature

In radio astronomy, the aerial or antenna temperature, T_{a}, is often used as an alternative way of expressing the power, P_{a}, of a celestial radio source incident upon the receiving element of a radio telescope (Lovell & Clegg, 1952). The antenna temperature may be related to the source power by:

$$T_{\mathrm{a}} = \frac{P_{\mathrm{a}}}{4k\,\Delta\nu} \tag{9.21}$$

where $\Delta\nu$ is the frequency bandwidth of the radio telescope and k is Boltzmann's constant.

9.4.4 Brightness temperature

The brightness temperature, T_{B}, is the temperature that a blackbody would need to have in order to match the observed specific intensity or spectral radiance of an astronomical source at an observed frequency ν, hence:

$$T_{\mathrm{B}} = \frac{c^2\,B_\nu}{2\,\nu^2\,k} \tag{9.22}$$

where the Rayleigh–Jeans approximation to the Planck law, valid for the low frequencies of radio astronomy, may be used.

If the celestial radio source is not a blackbody, such as a thin plasma, a maser or emitting synchrotron radiation, then the temperature derived using Equation (9.22) will not be correct and either too low or, sometimes, far too high.

9.5 Spectral classification as a temperature sequence

In Chapter 12, an outline of the way in which spectral types are assigned to stars using the relative strengths of the spectral lines produced by various ions, atoms and molecules is given. A relationship between temperature and spectral type or photometric colour differences may be determined empirically, which allows a base SI unit to be used to classify stars.

9.5.1 Energy states in atoms

Unlike, for example, the Solar System where an infinite number of orbits are possible for the planets, dwarf planets and other bodies such as comets, locations available to an electron relative to the nucleus of the atom of which it is a part is restricted by quantum theory to certain energy values. The absorption or emission of a photon by an electron allows it to move from one energy state to another, creating an absorption or emission line in the spectrum of the star.

The energy of a bound electron is given by Irwin (2007) as:

$$E_n = -\frac{2\pi^2}{n^2}\frac{m_e e^4}{h^2}\ \text{erg} \tag{9.23}$$

where m_e is the mass in grams and e is the charge of an electron in electrostatic units, h is Planck's constant in cgs units (erg . s) and n is the energy quantum level (an integer).

Substituting $m_e = 9.1093 \times 10^{-28}$ g, $e = 4.8032 \times 10^{-10}$ esu and $h = 6.626\,075 \times 10^{-27}$ erg . s in Equation (9.23) gives:

$$\begin{aligned} E_n &= \frac{-1}{n^2}(2.17987 \times 10^{-11})\,\text{erg} \\ &= \frac{-1}{n^2}(2.17987 \times 10^{-18})\,\text{J} \\ &= \frac{-1}{n^2}(13.61)\,\text{eV} \end{aligned} \tag{9.24}$$

Note that Equation (9.24) is given in cgs and esu units, resulting in a measure of energy also in cgs units, the erg. The transformation to SI units is accomplished by multiplying by the factor 10^{-7} since $1\,\text{J} = 10^7$ erg. The problems associated with transforming values in one set of electrostatic and electromagnetic units to another is dealt with in the following chapter (Chapter 10). Equation (9.24) also gives the transformation of the cgs energy value to one in electron volts.

The **electron volt** is a non-SI unit of **energy** accepted for use with the SI whose values in SI units are obtained experimentally. The electron volt dimension is $[\boldsymbol{M}].[\boldsymbol{L}]^2.[\boldsymbol{T}]^{-2}$ and symbol **eV**.

The electron volt is commonly used in spectroscopy. Its equivalent energy measurement in SI units is given by:

$$1\,\text{eV} = 1.602\,177\,33 \times 10^{-19}\,\text{J} \tag{9.25}$$

This numerical value is the same as the SI value for the electrical charge e of an electron, $1.602\,177\,33 \times 10^{-19}$ C (C = coulomb), which in turn is numerically the same as the esu value of the electron charge divided by $10\,c$ (c = speed of light = $2.997\,924$ m . s^{-1}), i.e.,

$$\begin{aligned} e &= 1.602\,177\,33 \times 10^{-19}\,\text{C} \\ &\equiv \frac{4.803\,2068 \times 10^{-10}\,\text{esu}}{10 \times 2.997924 \times 10^8\,\text{m.s}^{-1}} \end{aligned} \tag{9.26}$$

The capital C is the symbol for the SI unit of electrical charge, the coulomb (see Chapter 10).

9.5.2 Energy levels for the hydrogen atom

Using Equation (9.24) above it is possible to compute the values for the different energy levels ($n = 1, 2, 3$ etc. in the equation) for the hydrogen atom in SI units (J) as well as in eV. The energy difference $\Delta E_{j,k}$, between any two states j and k is given by:

$$\Delta E_{j,k} = E_j - E_k \simeq -2.18 \times 10^{-18} \left(\frac{1}{n_j^2} - \frac{1}{n_k^2}\right) \text{J} \tag{9.27}$$

$$\simeq -13.6 \left(\frac{1}{n_j^2} - \frac{1}{n_k^2}\right) \text{eV} \tag{9.28}$$

The frequency $\nu_{j,k}$, of the emitted or absorbed radiation is given by:

$$\nu_{j,k} = \frac{\Delta E_{j,k}}{h} \tag{9.29}$$

Example: find the energy, frequency and wavelength of the H_β spectrum line The lower level of the H_β Balmer line transition is at $n = 2$ and the upper level is at $n = 4$, so:

$$\Delta E_{2,4} = E_2 - E_4 \simeq -2.18 \times 10^{-18} \left(\frac{1}{2^2} - \frac{1}{4^2}\right)$$

$$= -4.088 \times 10^{-19} \text{J} \tag{9.30}$$

The frequency of H_β radiation is ν_{H_β} where:

$$\nu_{H_\beta} = \frac{\Delta E_{2,4}}{h} = 6.17 \times 10^{14} \text{Hz} \tag{9.31}$$

and its emission or absorption wavelength $\lambda_{H\beta}$ is:

$$\lambda_{H_\beta} = \frac{c}{\nu_{H_\beta}} = \frac{2.997\,924 \times 10^8}{6.17 \times 10^{14}} = 4.859 \times 10^{-7} \text{m}$$

$$= 485.9 \text{nm} \tag{9.32}$$

In Table 9.3, values for the energy differences, frequencies and wavelengths in SI units for the hydrogen Lyman, Balmer and Paschen line series are given up to energy level 6.

Grotrian diagram: Figure 9.2 is known as a Grotrian diagram. The numerical values for the energy levels in joules, computed above and listed in Table 9.3, have been plotted in Figure 9.2 as the ordinate. Each level is represented as an arbitrary 2 units along the abscissa. The arrow-headed lines show electron transitions from a lower to a higher state of energy representing the absorption of a photon; these are

Table 9.3. *Hydrogen spectrum lines*

Name	n	E_n J	E_n eV	$n_k - n_{1,2,3}$	ΔE J	ν Hz	λ nm
Lyman series							
L_∞	∞	0	0	$\infty - 1$	2.18 d − 18	3.29 d 15	91
L_ϵ	6	−6.06 d − 20	−0.82	6 − 1	2.12 d − 18	3.20 d 15	94
L_δ	5	−8.72 d − 20	−1.19	5 − 1	2.09 d − 18	3.16 d 15	95
L_γ	4	−1.36 d − 19	−1.85	4 − 1	2.04 d − 18	3.08 d 15	97
L_β	3	−2.42 d − 19	−3.30	3 − 1	1.94 d − 18	2.92 d 15	103
L_α	2	−5.45 d − 19	−7.41	2 − 1	1.63 d − 18	2.47 d 15	122
	1	−2.18 d − 18	−13.61				
Balmer series							
H_∞	∞	0	0	$\infty - 2$	5.45 d − 19	8.23 d 14	364
H_δ	6	−6.06 d − 20	−0.82	6 − 2	4.84 d − 19	7.31 d 14	410
H_γ	5	−8.72 d − 20	−1.19	5 − 2	4.58 d − 19	6.91 d 14	434
H_β	4	−1.36 d − 19	−1.85	4 − 2	4.09 d − 19	6.17 d 14	486
H_α	3	−2.42 d − 19	−3.30	3 − 2	3.03 d − 19	4.57 d 14	656
	2	−5.45 d − 19	−7.41				
Paschen series							
P_∞	∞	0	0	$\infty - 3$	2.42 d − 19	3.66 d 14	820
P_γ	6	−6.06 d − 20	−0.82	6 − 3	1.82 d − 19	2.74 d 14	1094
P_β	5	−8.72 d − 20	−1.19	5 − 3	1.55 d − 19	2.34 d 14	1282
P_α	4	−1.37 d − 19	−1.85	4 − 3	1.06 d − 18	1.60 d 14	1875
	3	−2.42 d − 19	−3.30				

In this table the shorthand notation $m\,\mathrm{d}\,n$ has been used where $m\,\mathrm{d}\,n \equiv m \times 10^n$.

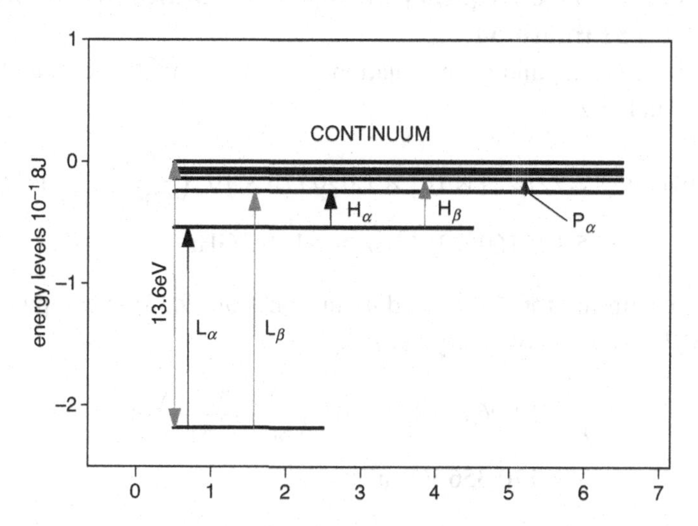

Figure 9.2. A Grotrian diagram for the Lyman (L), Balmer (H) and Paschen (P) series of hydrogen lines. The vertical axis is measured in units of 10^{-18} J. The energy difference in electron volts between the $n = 1$ and the $n = \infty$ is shown by the two-headed dark grey arrow labelled 13.6 eV.

known as bound-bound transitions. Transitions from a higher to a lower energy state occur with the emission of a photon; these are also bound-bound transitions. An arrow beginning at a horizontal line below the zero energy level and ending at a point above the zero energy level represents a bound-free transition. Free electrons (i.e., those not bound to an atomic, ionic or molecular nucleus) may undergo free-free transition in the region labelled 'CONTINUUM' in the figure.

The Rydberg formula: The following simple relationship between the inverse wavelength λ, in metres, of a spectrum line whose initial energy level is n_j and whose final energy level is n_k (for $n_j > n_k$) was discovered by Rydberg in the nineteenth century:

$$\frac{1}{\lambda} = R_{\mathrm{H}}\left(\frac{1}{n_j^2} - \frac{1}{n_k^2}\right)\mathrm{m}^{-1} \tag{9.33}$$

where R_{H}, the Rydberg constant in SI units, is for the hydrogen atom (Cox, 2000):

$$R_{\mathrm{H}} = 1.096\,775\,831 \times 10^7\,\mathrm{m}^{-1} \tag{9.34}$$

The inverse wavelength version of Rydberg's formula may be written in terms of frequency ν, and the speed of light c, as:

$$\nu = c\,R_{\mathrm{H}}\left(\frac{1}{n_j^2} - \frac{1}{n_k^2}\right)\mathrm{Hz} \tag{9.35}$$

Example: use the Rydberg formula to determine the frequency and wavelength of the radio-frequency hydrogen line formed by the energy level $n_j = 50$ to $n_k = 51$ transition

Substitute for c, R_{H}, n_j and n_k in Equation (9.35) to compute the value for the line frequency ν, in hertz:

$$\begin{aligned}\nu_{50-51} &= 2.997\,924 \times 10^8 \times 1.096\,776 \times 10^7\left(\frac{1}{50^2} - \frac{1}{51^2}\right)\mathrm{Hz} \\ &= 5.107\,160 \times 10^{10}\,\mathrm{Hz} \equiv 51.072\,\mathrm{GHz}\end{aligned} \tag{9.36}$$

Similarly, substituting for R_{H}, n_j and n_k in Equation (9.33) to compute firstly the value for λ^{-1} and then invert to yield λ:

$$\begin{aligned}\frac{1}{\lambda} &= 1.096\,775\,831 \times 10^7\left(\frac{1}{50^2} - \frac{1}{51^2}\right)\mathrm{m}^{-1} \\ &= 170.356\,569\,\mathrm{m}^{-1}\end{aligned} \tag{9.37}$$

and

$$\lambda = 5.870\,041 \times 10^{-3}\,\mathrm{m} \equiv 5.870\,\mathrm{mm} \tag{9.38}$$

9.5.3 The Saha–Boltzmann equation

The Saha–Boltzmann equation is valid when the condition of local thermodynamic equilibrium (LTE) prevails, i.e., assuming the behaviour of the matter component of the gas under consideration is governed by Maxwell–Boltzmann statistics, even if the electron component deviates from statistical equilibrium (see, e.g., Collins, 1989).

Under conditions of LTE, the relationship between the number density of atoms, N_i, in the ith state of ionization (i.e., the atom has had i electrons removed) to those atoms, N_{i+1}, in the $(i+1)$th ionization state, was shown by Saha (1921) to be:

$$\frac{N_{i+1}}{N_i} N_e = \frac{2u_{i+1}(T)}{u_i(T)} \frac{(2\pi m_e k T)^{\frac{3}{2}}}{h^3} \mathrm{e}^{-\frac{\chi_i}{kT}} \tag{9.39}$$

Equation (9.39) is sometimes written with electron pressure, P_e, substituted for N_e, the electron density, resulting in:

$$\frac{N_{i+1}}{N_i} P_e = \frac{2u_{i+1}(T)}{u_i(T)} \frac{(2\pi m_e)^{\frac{3}{2}}(kT)^{\frac{5}{2}}}{h^3} \mathrm{e}^{-\frac{\chi_i}{kT}} \tag{9.40}$$

where h is Planck's constant, k is Boltzmann's constant, m_e is the mass of the electron, χ_i is the ionization potential and is the energy required to move an electron from the ground state of an i-times ionized atom to the free (i.e., unbounded) state, T is the temperature of the gas and $u_i(T)$ is the partition function defined by:

$$u_i(T) = \sum_j g_j \mathrm{e}^{\frac{-\epsilon_j}{kT}} \tag{9.41}$$

where ϵ_j is the excitation potential above the ground state of the jth energy level of the atom or ion and g_j is the statistical weight ($g_j = 2J + 1$) of the jth energy level, J being the total angular quantum number of the energy level.

Example: calculate the ratio of singly ionized aluminium to neutral aluminium in a solar-type atmosphere at a point where the electron pressure is equal to 3 Pa

Using Equation (9.40) and substituting the following values from Cox (2000) and Aller (1963) (with transformation to SI units from cgs units where necessary) gives: $T = 5700\,\mathrm{K}$, $\chi_0 = 9.590\,265 \times 10^{-19}\,\mathrm{J}$, $\frac{2u_1}{u_0} = 0.34$, $P_e = 3\,\mathrm{Pa}$, $m_e = 9.1093 \times 10^{-31}\,\mathrm{kg}, h = 6.626075 \times 10^{-34}\,\mathrm{J.s}, k = 1.380658 \times 10^{-23}\,\mathrm{J.K}^{-1}$

$$\frac{N_1}{N_0} = \frac{2u_1(T)}{u_0(T)} \frac{(2\pi m_e)^{\frac{3}{2}})(kT)^{\frac{5}{2}}}{P_e h^3} \mathrm{e}^{-\frac{\chi_0}{kT}} \tag{9.42}$$

$$= 47.27 \tag{9.43}$$

i.e.,

$$N_1 = 47.27\, N_0 \tag{9.44}$$

so just over 2 % of the aluminium atoms are neutral and the rest are singly ionized.

9.5.4 Equivalent width of a spectral line

The profile of a spectral line formed in the atmosphere of a star is dependent on many different variables, such as the atmospheric pressure, the temperature, the rotation velocity of the star, the effects of any electric or magnetic fields present and so on. The general shape may be described as Gaussian, unless the line is saturated. The measured area contained within the line profile may be set equal to that of a rectangular area, whose height is equal to that of an absorption line that is completely black at the centre and whose width is then assigned a value such that the profile area and the rectangular area match (see Figure 9.3). This width is known as the **equivalent width**, W_λ, of the spectral line and is defined in wavelength space to be:

$$W_\lambda = \int_{\lambda_{min}}^{\lambda_{max}} \left(\frac{1 - F_\lambda}{F_0} \right) d\lambda = F_0 \, . \, (\lambda_2 - \lambda_1) \tag{9.45}$$

where F_0 is the flux density of the continuum fitted to the observed stellar spectrum and F_λ is the flux density at wavelength λ over the wavelength range of interest ($\lambda_{min} \rightarrow \lambda_{max}$). Collins (1989) notes that for narrow spectral lines with ($\lambda_{max} - \lambda_{min}) \ll \lambda_0$ it is possible to write the equivalent width, W_ν, in frequency space as:

$$W_\nu \approx \frac{W_\lambda \, \nu_0}{\lambda_0} \tag{9.46}$$

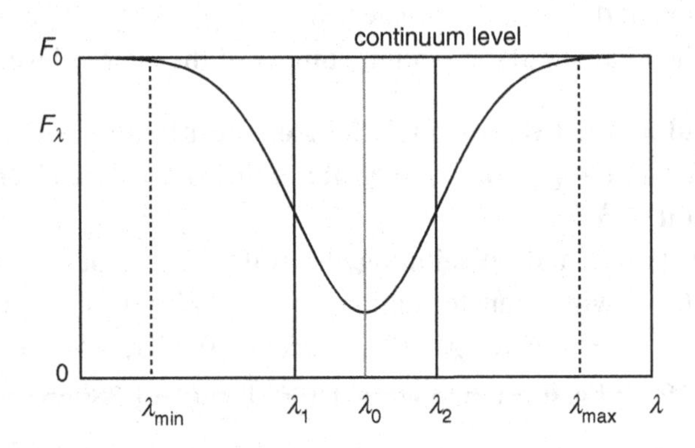

Figure 9.3. The area enclosed by the Gaussian-shaped spectral line profile between λ_{min} and λ_{max}, centred on wavelength λ_0 is equal to the rectangular area, $F_0 \, . \, (\lambda_2 - \lambda_1)$, whose width, $(\lambda_2 - \lambda_1)$, is defined to be the equivalent width of the actual spectral line.

By assigning appropriate values to the constants and variables in Equations (9.39), (9.40) and (9.41) it is possible to derive the ratio of spectral line strengths for selected atoms and ionized atoms. This ratio is a function of the effective temperature of the star. The relative strengths of spectrum lines in stars may be directly measured from stellar spectrograms and thus used in conjunction with the line-strength ratios determined from the Saha–Boltzmann equation to produce an estimate of the star's temperature. Since the ratios of line strengths are also used to classify stellar spectra (see Chapter 12) and determine spectral class, it becomes possible to equate spectral class directly with stellar temperature in kelvin (see Figure 9.4). More detailed discussions on the use of the Saha–Boltzmann equation may be found in, e.g., Lang (2006), Irwin (2007), Collins (1989), Cox (2000) and Aller (1963).

9.5.5 Colour temperature

For unreddened or dereddened monochromatic flux density measurements of stars obtained via multicolour photometry, a direct relationship exists between spectral type and colour index and hence between temperature and colour index.

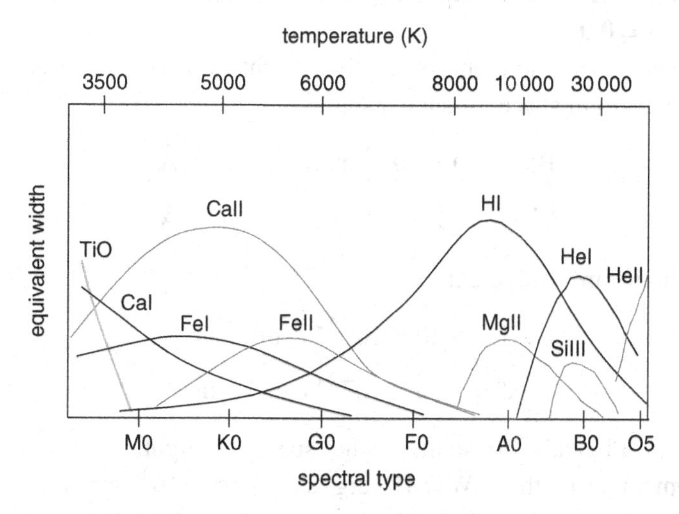

Figure 9.4 A plot showing the variation of the strength (equivalent width) of various atoms (e.g., FeI iron) and ions (e.g., FeII singly ionized iron) against both spectral type and temperature. This figure was adapted, reversed and updated from the original, which appeared in Struve *et al.* (1959). The black lines are for neutral atoms, the dark grey lines are for ions and the wider light grey line is for the molecule TiO. It can readily be seen that the ratio of equivalent widths of atoms and/or ions to other atoms and/or ions is a function of spectral type and temperature.

The colour temperature of an astronomical body may be determined if the relative energy distribution of the body is known. Consider the apparent monochromatic flux densities f_1 and f_2, at two wavelengths λ_1 and λ_2, then the ratio of f_1 and f_2 is the same as that derived using Planck's law, so:

$$\frac{f_1}{f_2} = \frac{\lambda_2^5}{\lambda_1^5} \frac{e^{\frac{hc}{\lambda_2 kT}} - 1}{e^{\frac{hc}{\lambda_1 kT}} - 1} \tag{9.47}$$

where c is the speed of light, h is Planck's constant, k is Boltzmann's constant and T is the colour temperature. Rewriting the flux values as Pogson magnitudes, m_1 and m_2, gives:

$$m_1 - m_2 = -2.5 \log\left(\frac{f_1}{f_2}\right) + c_1 \tag{9.48}$$

where constant c_1, depends on the different zero points of the magnitude scales used. The Wien approximation may be used in the optical part of the spectrum to allow the colour temperature, T_c, to be written as:

$$T_c = \frac{2.5 \log(e)\, h\, c \left(\frac{1}{\lambda_1} - \frac{1}{\lambda_2}\right)}{k\left((\mathrm{B} - \mathrm{V}) - c_1 + 2.5 \log\left(\frac{\lambda_2}{\lambda_1}\right)^5\right)} \tag{9.49}$$

Example: determine the colour temperature of an A0V star with colour index (B – V) = 0.0

Cox (2000) gives the monochromatic flux densities in cgs units in the B and V bands, in wavelength space, of such a star as:

$$\mathrm{B} = 6.40 \times 10^{-9}\,\mathrm{erg\,.\,s^{-1}\,.\,cm^{-2}\,.\,\AA^{-1}}$$
$$\mathrm{V} = 3.75 \times 10^{-9}\,\mathrm{erg\,.\,s^{-1}\,.\,cm^{-2}\,.\,\AA^{-1}} \tag{9.50}$$

which in SI units are equivalent to:

$$\mathrm{B} = 6.40 \times 10^{-2}\,\mathrm{W\,.\,m^{-2}\,.\,m^{-1}}$$
$$\mathrm{V} = 3.75 \times 10^{-2}\,\mathrm{W\,.\,m^{-2}\,.\,m^{-1}} \tag{9.51}$$

with h being Planck's constant, c the speed of light, k Boltzmann's constant and remembering that $1\,\mathrm{W} = 10^7\,\mathrm{erg\,.\,s^{-1}}$, $1\,\mathrm{m}^2 = 10^4\,\mathrm{cm}^2$ and $1\,\mathrm{m} = 10^{10}\,\mathrm{\AA}$. Substituting the appropriate values into Equations (9.48) and (9.49) gives:

$$(\mathrm{B} - \mathrm{V}) = 0 = -2.5 \log\left(\frac{3.75 \times 10^{-2}}{6.40 \times 10^{-2}}\right) + c_1 \tag{9.52}$$

rearrange for c_1:

$$c_1 = -0.58 \tag{9.53}$$

and for T_c:

$$T_c = 12340\,\text{K} \tag{9.54}$$

Note that this value is considerably higher than the value for the effective temperature of 9790 K given by Cox (2000). Selecting other pairs of photometric bands would produce different values for the colour temperature. By way of illustration, consider the red and near-infrared bands R and I.

Example: determine the colour temperature of an A0V star with colour index (R – I) = 0.0

The computation is as for the previous example, substituting the appropriate values for R and I wavelengths in metres and monochromatic flux densities in $\text{W}\,.\,\text{m}^{-2}\,.\,\text{m}^{-1}$ into Equations (9.48) and (9.49).

$$\lambda_{\text{R}} = 7.1 \times 10^{-7}\,\text{m} \qquad f_{\text{R}} = 1.75 \times 10^{-2}\,\text{W}\,.\,\text{m}^{-2}\,.\,\text{m}^{-1}$$
$$\lambda_{\text{I}} = 9.7 \times 10^{-7}\,\text{m} \qquad f_{\text{I}} = 8.40 \times 10^{-3}\,\text{W}\,.\,\text{m}^{-2}\,.\,\text{m}^{-1}$$
$$(\text{R} - \text{I}) = 0.0 \tag{9.55}$$

The colour temperature so derived is:

$$T_c = 7398\,\text{K} \tag{9.56}$$

The colour temperature does not represent the physical temperature within the photosphere of a star, but rather may be used to give an indication of the slope of the energy distribution emitted by a star.

9.5.6 A lookup table to convert (B – V) colour index to effective temperature

A **lookup table** is defined in computer science to be a data structure such as a numerical or alphanumerical array that may be used to replace a runtime computation. From a listing such as that given in Cox (2000) of effective temperature against spectral type and various colour indices, it is possible to use linear interpolation to estimate a value for the temperature of a star. The listed colour indices are dereddened so that any observational colour index must first be corrected for the effects of interstellar obscuration due to dust and gas. From the lookup table, extract the values and colour index and corresponding effective temperature immediately above $(\text{B} - \text{V})_i$ and below $(\text{B} - \text{V})_{i+1}$ the observed star's dereddened colour index $(\text{B} - \text{V})_\star$. The interpolated effective temperature of the star, $T_{\text{eff},\star}$, is then calculated from:

$$T_{\text{eff},\star} = \left(\frac{T_{\text{eff},i+1} - T_{\text{eff},i}}{(\text{B} - \text{V})_{i+1} - (\text{B} - \text{V})_i}\right)(\text{B} - \text{V})_\star + \kappa \tag{9.57}$$

where

$$\kappa = T_{\mathrm{eff},i} + \left(\frac{T_{\mathrm{eff},i+1} - T_{\mathrm{eff},i}}{(\mathrm{B-V})_{i+1} - (\mathrm{B-V})_i}\right)(\mathrm{B-V})_i \tag{9.58}$$

Example: determine the effective temperature of the Sun from a lookup table (Cox, 2000)

Substitute the following data into Equations (9.57) and (9.58):

$$T_{\mathrm{eff},i} = 5790\,\mathrm{K} \qquad T_{\mathrm{eff},i+1} = 5560\,\mathrm{K}$$
$$(\mathrm{B-V})_i = 0.68 \qquad (\mathrm{B-V})_{i+1} = 0.63$$
$$(\mathrm{B-V})_{\odot} = +0.650$$

to give:

$$\kappa = 5790 + \left(\frac{5560 - 5790}{0.68 - 0.63}\right)(0.63) = 8688\,\mathrm{K}$$

$$T_{\mathrm{eff},\odot} = \left(\frac{5560 - 5790}{0.68 - 0.63}\right)(0.650) + 8688 = 5698\,\mathrm{K}$$

which may be compared with the value of 5777 K for the effective temperature of the Sun given in Cox (2000). A plot of effective temperature against (B – V) colour index is given in Figure 9.5. The location of the Sun in the plot is shown with an arrow.

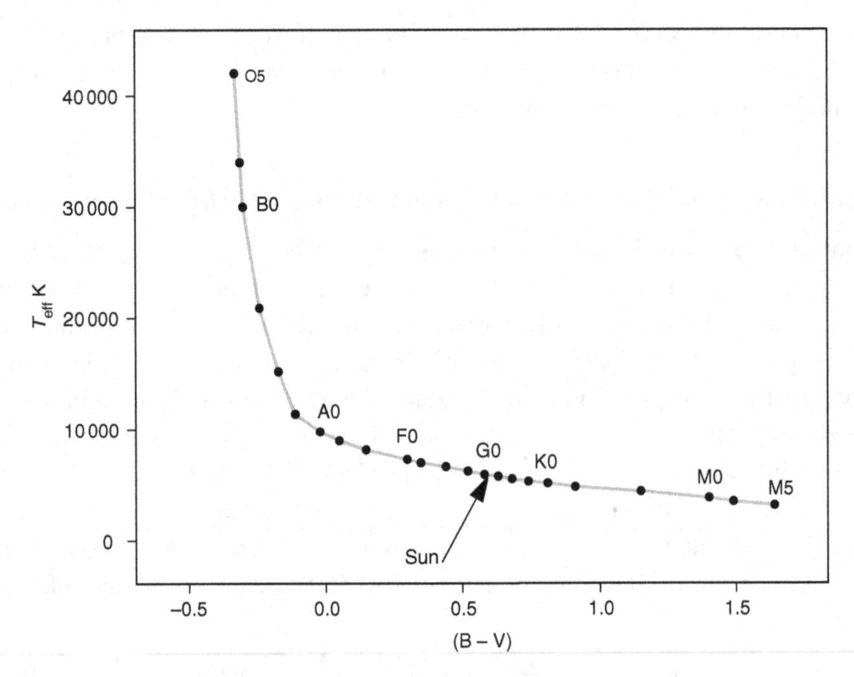

Figure 9.5. A colour index (B – V), versus (T_{eff}) for main sequence stars from type O5 to M5 (Cox, 2000).

So using the spectrophotometric or lookup table methods outlined above it is possible to replace spectral type and colour index by an approximate effective temperature in SI unit kelvin, but the use of colour temperature rather than colour index would generally be a poor substitute and at worst even misleading.

9.6 Model stellar atmospheres

Currently, the most realistic way of determining stellar photospheric temperatures is to compare either an observational spectrogram or a set of discrete photometric measurements with a computer-generated grid of model stellar atmospheres.

A **model stellar atmosphere** may be defined as a mathematical construct or description of a real stellar atmosphere, with specified input temperature, pressure, density and chemical composition, that obeys a set of fundamental physical laws.

The simplest and most common models assume that there are no magnetic or electric fields present, no bulk movements in the atmosphere (e.g., convection or rotation), the ideal gas laws are obeyed and the atmosphere, (generally considered to be plane parallel or spherically symmetrical) is in local thermodynamic equilibrium and hydrostatic equilibrium.

The data input to the models include the effective temperature, the gravitational acceleration at the surface and the chemical composition of the atmosphere, normally cited relative to that of the solar atmosphere. The model which best fits the observed emergent stellar flux against wavelength or frequency values is taken to give a reasonable indication of the most likely actual values of effective temperature, surface gravity and chemical composition. If the resolution of the grid spacing is too coarse then the use of a subset of models to allow a refined interpolated group of values should be used.

9.6.1 Kurucz model stellar atmospheres

A major grid of model stellar atmospheres was published by Kurucz (1979) for stars of spectral type O, B, A, F and G with effective temperatures from 5500 K to 50 000 K and chemical abundances 0.01 solar, 0.1 solar and solar values. An update and extension of this work in 1993 by Kurucz[59] was made to include stars of later spectral type and hence lower surface temperature (down to 3000 K). The wavelength range runs from 9 nm to 160 μm with a grid step from 1 nm at the short-wavelength end to 40 nm at the long. The revised chemical abundance relative to the Sun ($\log Z$) varies from $\log Z = +1.0$ (i.e., 10 times the solar abundance) to $\log Z = -5.0$ (0.000 01 times solar abundance). Surface gravity ($\log g$) varies in the range $\log g = 0.0\,(0.5)\,5.0$.

[59] See http://www.stsci.edu/hst/observatory/cdbs/k93models.html

The output of the Kurucz models is in the format (λ, f_λ), where λ is in nm and F_λ is in erg . s^{-1} . cm^{-2} . Hz^{-1} . sr^{-1}.

9.6.2 Conversion of Kurucz model output from cgs to SI units

The output of wavelengths is either in nm or Å, so the conversion to SI units in m is achieved by dividing by 10^9 or 10^{10}. The flux output unit conversion is a little more complex, the first stage being to convert the component parts of the composite unit from cgs to SI form.

$$\begin{aligned} \mathrm{erg\,.\,s^{-1}\,.\,cm^{-2}\,.\,Hz^{-1}\,.\,sr^{-1}} &= (10^{-7})\,\mathrm{J\,.\,s^{-1}\,.\,(10^{-2})^{-2}\,m^{-2}\,.\,Hz^{-1}\,.\,sr^{-1}} \\ &= 10^{-3}\,\mathrm{W\,.\,m^{-2}\,.\,Hz^{-1}\,.\,sr^{-1}} \\ &= 10^{23}\,\mathrm{Jy\,.\,sr^{-1}} \end{aligned} \tag{9.59}$$

The total emergent monochromatic flux from the star is 4π times the unit steradian value above. If the radius of the emitting star is $\rho_\star$ metres and its distance is $R_\star$, then the conversion factor, conv, between Kurucz units and janskys becomes:

$$\mathrm{conv} = 4\pi \left(\frac{\rho_\star}{R_\star}\right)^2 \times 10^{23} \tag{9.60}$$

or if f_J is the monochromatic flux density in janskys and f_K is the monochromatic flux density in Kurucz units, then:

$$f_\mathrm{J} = 4\pi \left(\frac{\rho_\star}{R_\star}\right)^2 . \, 10^{23} \times f_\mathrm{K} \tag{9.61}$$

Determination of values for a star's radius and its distance

As an observational determination of the radius of a star is at present not generally available for the majority of stars, those that do have such measures have been used to calibrate a relationship between spectral type and radius. For main sequence stars, the values of radius vary from 1.029×10^{10} m for spectral type O5 to 1.325×10^8 m for spectral type M5 according to Straižys (1992). In Table 9.4, the conversion from solar radii to SI units was achieved using a value for the solar radius of $R_\odot = 9.955\,08 \times 10^8$ m (Cox, 2000).

The determination of stellar distances is dealt with in Chapter 6.

Example: plot the comparison between the absolute HST spectrum of Vega and a Kurucz model stellar atmosphere

By the absolute HST spectrum of Vega is meant the unreddened spectrum that would be observed at a distance of 10 pc or $3.085\,677\,6 \times 10^{17}$ m. The model stellar

Table 9.4. *Stellar radii in solar radius and SI units against spectral class and luminosity class*

Spectral luminosity class	Radius $R_\odot$ m	Spectral luminosity class	Radius $R_\odot$ m	Spectral luminosity class	Radius $R_\odot$ m
O5V	14.791	O5III	17.783	O5Iab	22.909
	1.029 d 10		1.237 d 10		1.593 d 10
B0V	7.244	B0III	10.965	B0Iab	25.119
	5.039 d 9		7.626 d 9		1.747 d 10
A0V	2.291	A0III	3.631	A0Iab	63.096
	1.593 d 9		2.525 d 9		4.388 d 10
F0V	1.413	F0III	2.570	F0Iab	114.82
	9.824 d 8		1.788 d 9		7.986 d 10
G0V	1.072	G0III	–	G0Iab	245.47
	7.453 d 8		–		1.707 d 11
K0V	0.977	K0III	10.00	K0Iab	389.05
	6.797 d 8		6.955 d 9		2.706 d 11
M0V	0.603	M0III	43.65	M0Iab	524.81
	4.191 d 8		3.036 d 10		3.650 d 11
M5V	0.191	M5III	109.6	M5Iab	–
	1.325 d 8		7.626 d 10		–

In this table the shorthand notation $m\,\mathrm{d}\,n$ has been used where $m\,\mathrm{d}\,n \equiv m \times 10^n$.

atmosphere is also computed as though the star were at a distance of 10 pc. The Kurucz model selected has an effective temperature of 9750 K, a surface gravity of $\log g = 4.00$ and solar chemical abundances ($\log Z = 0.00$). The radius of an A0V star is, from Table 9.4, 1.593×10^9 m. Using the conversion between Kurucz units and janskys given in Equation (9.61) and inserting the values for the stellar radius and its distance, plus a value for the monochromatic flux density of $4.93 \times 10^6\ \mathrm{erg.s^{-1}.cm^{-2}.Hz^{-1}.sr^{-1}}$ from the Kurucz model at a wavelength of 3 μm (or a frequency of 10^{14} Hz) gives:

$$f_{\mathrm{J}} = 4\pi \left(\frac{1.593 \times 10^9}{3.0856776 \times 10^{17}}\right)^2 \times 10^{23} \times 4.93 \times 10^6$$
$$= 165.03\ \mathrm{Jy} \tag{9.62}$$

Figure 9.6 shows segments of the observed and theoretical spectra from 10^{14} Hz to approximately 2×10^{15} Hz.

9.6.3 Mean monochromatic flux densities from Kurucz models

The Kurucz model stellar atmospheres are essentially step functions, so computing the mean monochromatic flux density output, f_ν, in a given frequency band is

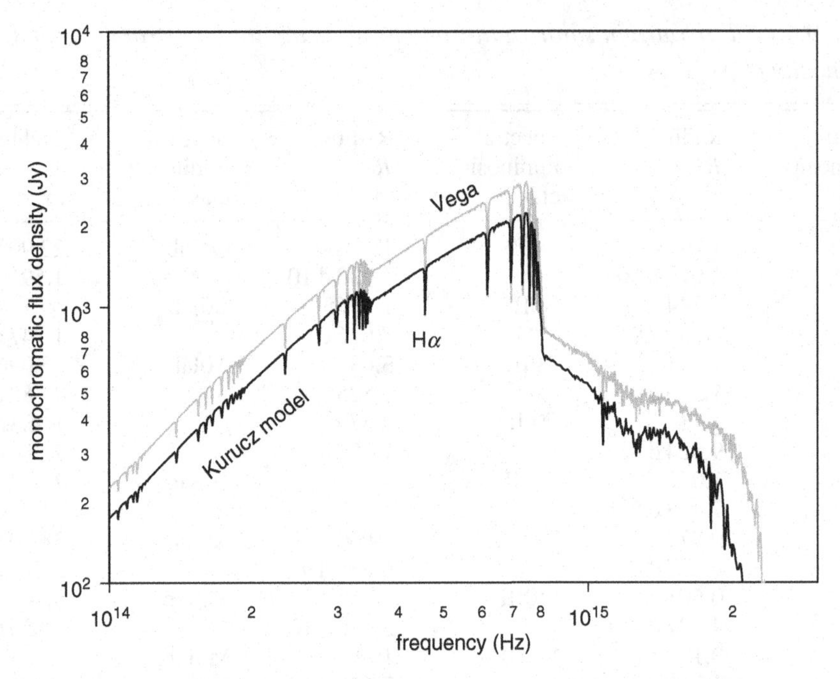

Figure 9.6. Comparison between the HST spectrum of the A0V star Vega and a Kurucz model stellar atmosphere, with parameters $T = 9\,750$ K, $\log g = 4.00$ and $\log Z = 0.00$. The effective temperature was deliberately set too low to allow for easier comparison between the model-generated spectrum lines and the actual HST spectrum. Note that the monochromatic flux densities of Vega and the Kurucz model have been adjusted to appear as though they were both at the standard distance of 10 pc.

simply a matter of summing the product of the flux elements ϕ_ν and the elemental frequency bandwidths δ_ν and dividing by the sum of the frequency bandwidth elements, i.e.,

$$f_\nu = \frac{\sum \phi_\nu \delta_\nu}{\sum \delta_\nu} \tag{9.63}$$

Since, in general, the maximum and minimum frequency values lie between tabulated model frequencies, end corrections using simple linear interpolation were made. Monochromatic flux densities were calculated for various photometric bands from the satellite ultraviolet (the ANS and TD1 satellites), through the optical spectrum (Vilnius and Johnson photometry), to the infrared (USNO-B I band, 2MASS and IRAS photometry). The number of model data points available per band varied from 1 for the IRAS 60μm and 100μm bands to a maximum of 95 for the

Table 9.5. *Wavelength frequency and mean monochromatic flux density data for Kurucz model (T_{eff} = 9750 K, log g = 4.00, log Z = 0.00) photometric bands*

Band ID	$\overline{\lambda}$ nm	$\Delta\lambda$ nm	$\overline{\nu}$ THz	$\Delta\nu$ THz	F_ν Jy	n_K
ANS1	154	5	1940	63	206.4	7
ANS2	154	14	1935	187	215.9	18
ANS3	179	14	1666	138	327.9	17
ANS4	220	20	1363	124	390.0	22
ANS5	249	15	1203	72	397.7	18
ANS6	329	10	910	28	623.4	7
TD1	156	33	1916	409	216.1	38
TD2	196	33	1526	258	367.6	36
TD3	236	33	1268	178	382.4	36
TD4	274	31	1094	124	466.2	32
U	345	40	869	101	650.3	23
P	374	26	802	56	968.4	15
X	405	22	740	40	1922.5	13
Y	466	26	643	36	2005.0	15
Z	516	21	581	24	1809.1	12
V	544	26	551	26	1718.5	14
S	656	20	457	14	1322.9	12
U	367	66	817	148	955.6	36
B	436	94	688	150	1980.5	48
V	545	88	550	89	1728.5	46
I	808	185	371	86	1132.5	95
J	1235	162	243	32	725.9	34
H	1662	251	180	28	459.3	34
K_S	2159	262	139	17	302.7	28
L	3600	1200	83	29	126.9	71
M	4800	800	63	11	71.2	42
12 μm	12 000	7000	25	16	15.1	42
25 μm	25 000	11 150	12	6	3.0	3
60 μm	60 000	32 500	5	3	0.5	1
100 μm	100 000	31 500	3	1	0.2	1

USNO-B I band. Table 9.5 lists the band identification, mean wavelength $\overline{\lambda}$, bandwidth $\Delta\lambda$, mean band frequency $\overline{\nu}$, frequency bandwidth $\Delta\nu$, the mean absolute monochromatic flux density, F_ν, emitted by the model star atmosphere at a distance of 10 pc, and number n_K of Kurucz model data points used in computing the mean monochromatic flux density. Wavelength units are nm and frequency units THz (10^{12} Hz).

Figure 9.7 illustrates the output for the USNO-B I band of the Kurucz model stellar atmosphere with T_{eff} = 9750 K, log g = 4.00 and log Z = 0.00. The model

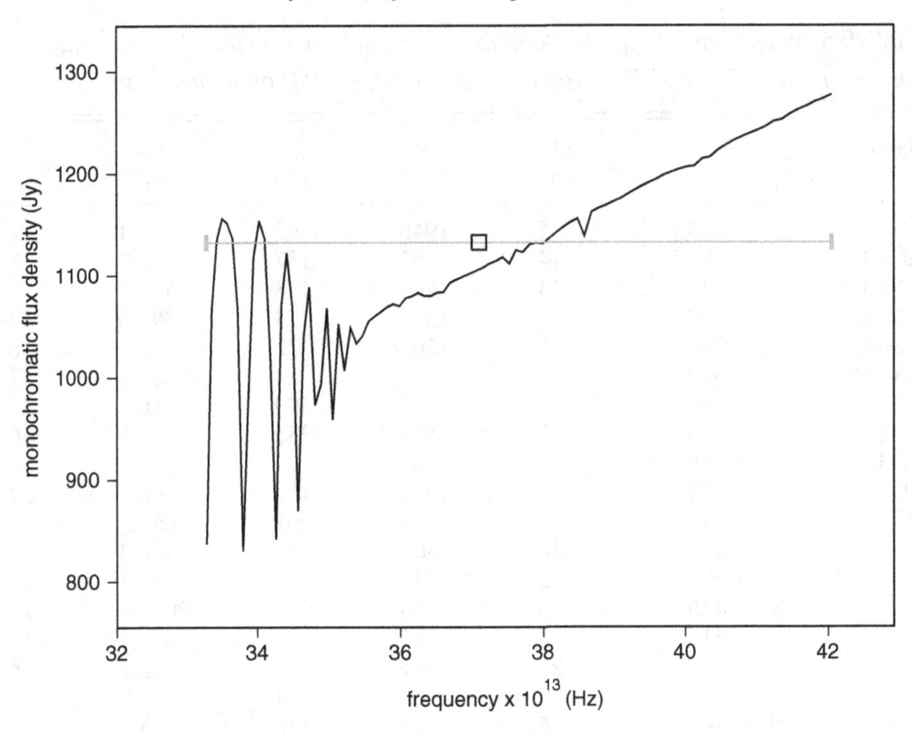

Figure 9.7. USNO-B I band mean monochromatic flux density and frequency bandwidth against Kurucz model stellar atmosphere output. The $\nu_{\rm min}$ and $\nu_{\rm max}$ values for each monochromatic flux density sample computed are shown, which creates the step-function-like appearance. The construction of the I band used 95 sample fluxes from the Kurucz model for $T = 9750\,\rm K$, $\log g = 4.00$ and $\log Z = 0.00$. The light-grey line shows the mean monochromatic flux measured in the bandwidth shown by the horizontal light-grey line. The black square is located at the mean frequency and mean monochromatic flux density of the band.

star is taken to be at a distance of 10 pc (3.085×10^{17} m). The single light-grey line plotted shows the frequency bandwidth and the mean value (the black square) of the monochromatic flux density in janskys. The prominent line formation at the low-frequency end of the band is due to the Paschen series of hydrogen.

By way of an example, a plot is shown in Figure 9.8 comparing the observed photometry of the star HD73952, a member of the young open star cluster IC2391 (Dodd, 2007), with two Kurucz model stellar atmospheres, with the input parameters $\log g = 4.00$, $\log Z = 0.00$ and $T_{\rm eff} = 9750\,\rm K$ and $T_{\rm eff} = 11\,750\,\rm K$. The respective monochromatic flux density outputs have been adjusted to be those that would be observed were the star (and the models) at a distance of 10 pc and dereddened. The points shown with error bars are those derived from observation, the point itself being at the mean frequency and the mean monochromatic flux

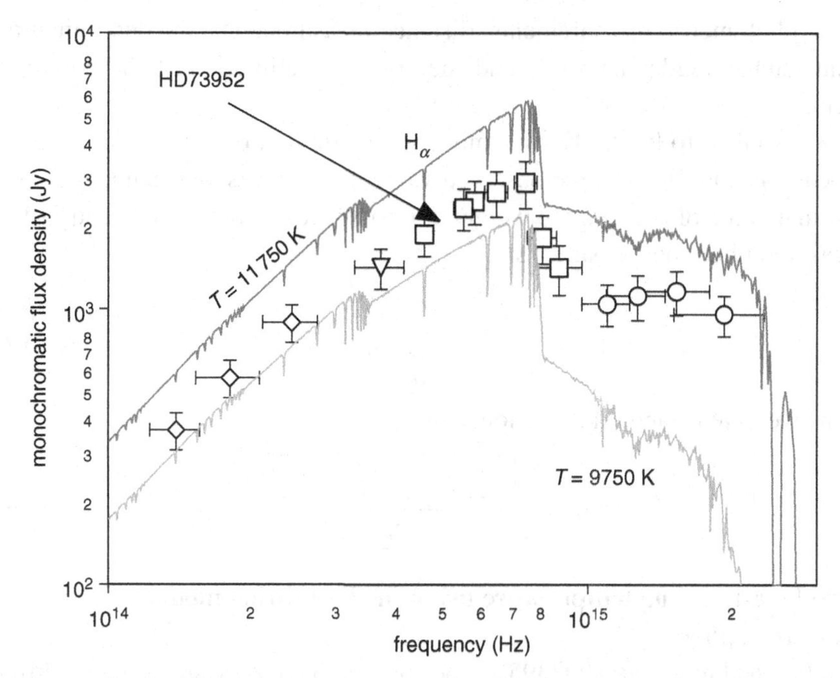

Figure 9.8. A comparison between two computed model stellar atmospheres, with $\log g = 4.00$, $\log Z = 0.00$ and temperatures 9750 K and 11 750 K, with actual photometric observations of the B8Vn star HD73952. The error bars attached to each observational point represent the photometric bandwidth and an estimate of the total error in the photometry. Circles are TD1 photometry, squares are Vilnius photometry, the inverted triangle is the USNO-B I band and the diamonds 2MASS photometry. The monochromatic flux densities in janskys are adjusted to those that would be observed were the models and the star at the standard 10 pc distance.

density. The error bar in the x direction being the frequency bandwidth and that in the y direction being due to the uncertainty in the observational photometric measurement.

9.6.4 Selection of best fitting model

Computing the weighted mean and weighted standard deviation of the differences between the Kurucz model monochromatic flux density in a given photometric band and the observational value was not used, given the wide variation in the monochromatic flux density with frequency that could readily bias the mean towards the difference value of the band with the greatest signal. Instead, the weighted mean of the ratios of the observational to model monochromatic flux densities were used. The weighting applied was the ratio of the measured average signal strength to the standard deviation about that mean due to contributing variables, such as photometric measurements, distance measurements, the determination of the interstellar reddening value and the absolute calibration of the photometric system.

If K_i is taken to be the Kurucz monochromatic flux density at the mean band frequency ν_i and O_i to be the observed value with a measured standard deviation about that value of σ_i, then the weighted mean ratio, R, between K and O for all n measured photometric bands is:

$$R = \frac{\sum_{i=1}^{n} \frac{O_i^2}{\sigma_i K_i}}{\sum_{i=1}^{n} \frac{O_i}{\sigma_i}} \tag{9.64}$$

and the weighted standard deviation, σ_R is:

$$\sigma_R = \sqrt{\left[\frac{\sum_{i=1}^{n} \frac{O_i}{\sigma_i}(\frac{O_i}{K_i} - R)^2}{\frac{n-1}{n}\sum_{i=1}^{n} \frac{O_i}{\sigma_i}}\right]} \tag{9.65}$$

Example: estimating temperature using the best fitting model stellar atmosphere

Using this technique on HD73952, a set of weighted χ^2 values against different effective temperatures of Kurucz models with the same values of $\log g$ and $\log \mathrm{Z}$ may be determined. A second-order fit to the curve is then made and the minimum value of weighted χ^2 parameter used to find the corresponding value of T_{eff}.

In Figure 9.9 this may be seen to be approximately 10 900 K. Straižys (1992) assigns that temperature to a star of type B8–9V, which would seem a good start in determining a more precise temperature value using the model stellar atmospheres in an iterative manner. (SIMBAD lists HD73952 to be spectral type B8Vn.)

9.7 Summary and recommendations

9.7.1 Summary

The SI unit of thermodynamic temperature, the kelvin, has been developed and refined in parallel with other, non-absolute scales, such as those of Celsius and Fahrenheit and one other absolute scale due to Rankine. Conversion formulae between these systems are given, with examples. Sample temperatures of various astronomical locations from the surface of the Earth to the cosmic background radiation due to the Big Bang are given in kelvin.

The many different types of temperature, such as thermodynamic, kinetic, effective, Planck, blackbody, antenna, brightness and colour are defined and worked examples of temperatures derived using appropriate observational techniques (radiometry, photometry, spectrophotometry) given. The possibility of using a temperature sequence in place of spectral classification is discussed in some detail.

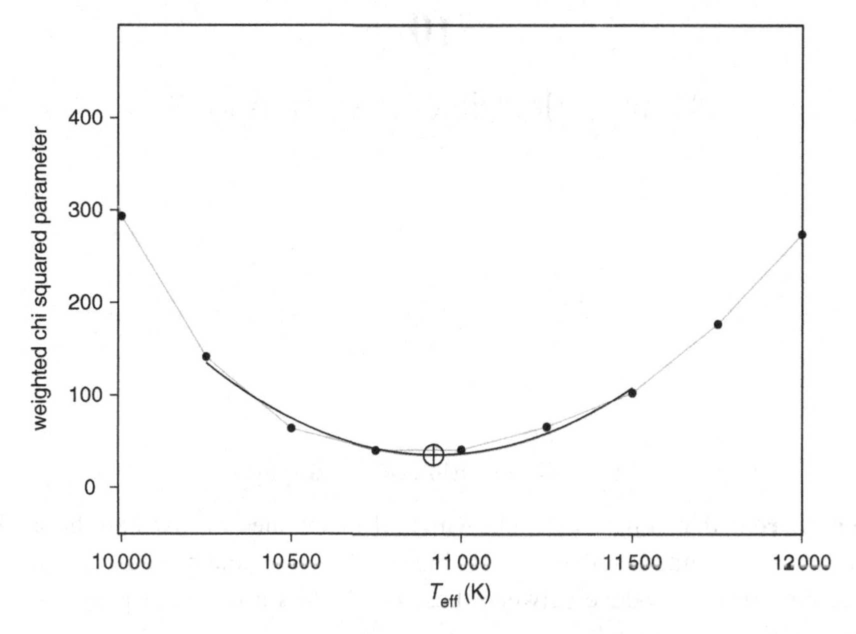

Figure 9.9. A plot of a weighted χ^2 parameter against effective temperatures from Kurucz model stellar atmospheres. The black points are actual values, the grey lines merely link up the points and the solid black line is a second-order least squares fit between the two variables. The minimum value of the weighted χ^2 parameter corresponds to an effective temperature of 10 919 K.

The final section illustrates how model stellar atmospheres in conjunction with coarse (broadband and intermediate-band photometry) and standard spectrophotometry may be used to calculate stellar temperatures.

9.7.2 Recommendations

The basic recommendation to most astronomers would be to carry on with what they are doing now, as most already use the SI unit, the kelvin, as a measure of temperature.

Some popular astronomical publications often cite temperatures in degrees Celsius or Fahrenheit and should be encouraged to show the equivalent temperature in kelvin in parenthesis. This would lead to an increasing familiarity with kelvin amongst the interested laity.

Whilst it is possible to replace spectral class and dereddened colour indices by a measure of temperature, there would seem to be no really good reason for doing so, though it may be argued that for ease of comparison between observations and theory, similar variables should be used.

10

Unit of electric current (ampere)

10.1 SI definition of the ampere

The **ampere** is that constant current which, if maintained in two straight parallel conductors of infinite length, of negligible cross section, and placed 1 metre apart in vacuum, would produce between these conductors a force equal to 2×10^{-7} newton per metre of length.

The dimension of electric current is [$\boldsymbol{I}$], its unit is the **ampere** and its symbol is **A**.

10.1.1 Possible future definition of the ampere

Under discussion at the present time is the redefining of the unit of electric current in the following way:

The ampere is a unit of electric current such that the elementary charge is exactly $1.602\,176\,53 \times 10^{-19}$ C, where 1 C (coulomb) = 1 A . s (ampere second).

10.1.2 Definition of electricity

Funk & Wagnalls New Standard Dictionary of the English Language (1946) gives a general definition of electricity as:

Electricity: A material agency which, when in motion, exhibits magnetic, chemical and thermal effects, and which, whether in motion or at rest, is of such a nature that when it is present in two or more localities within certain limits of association, a mutual interaction of force between such localities is observed.

And as a physics-related definition of electricity:

That branch of science that treats of this agency and the phenomena caused by it.

10.1.3 Definition of magnetism

Funk *et al.* (1946) give, as a general definition of magnetism:

Magnetism: that quality or agency by virtue of which certain bodies are productive of magnetic force or susceptible to its action.

And as a physics-related defintion of magnetism:

The science that treats the laws and conditions of magnetic force.

10.2 SI and non-SI electrical and magnetic unit relationships

There are five major systems of electrical and magnetic units in use at present, namely: the electrostatic system (esu), the electromagnetic system (emu), the cgs Gaussian, the practical system and the SI system. Of these, the first three are based on the cgs system for all mechanical quantities, but differ in that the esu system is additionally based on the law of force between electric charges, whilst the emu system is based on the corresponding law of force between magnetic poles with the unrationalized cgs Gaussian being a hybrid using both esu and emu units. A difficulty arises in everyday use of either the esu or emu system in that particular units are considered to be either too small or too large, so a further system, the practical system, was developed. The practical system is based on the metre, kilogram, second and ampere, similar to, but not identical with, the SI. The SI naturally uses the metre, kilogram and second as the fundamental mechanical units, in conjunction with the ampere as a fourth unit to define electrical and magnetic quantities.

The IAU recommends the use of the SI units only and particularly deprecates (Wilkins, 1989) the use of the gauss or the gamma for magnetic flux densities and the oersted for magnetic field strength. Burke & Graham-Smith (2002) state in a footnote on page 107 of their radio astronomy textbook: 'The usual unit in astrophysics (for magnetic fields): 1 gauss $= 10^{-4}$ tesla.' The rest of this section will therefore consider mainly the relationship between the unrationalized cgs Gaussian and the SI unit systems.

SI and unrationalized cgs Gaussian base units

Electric and magnetic derived units in the SI system are constructed from the four base units: the kilogram, the metre, the second and the ampere, with dimensions $[\boldsymbol{M}]$, $[\boldsymbol{L}]$, $[\boldsymbol{T}]$ and $[\boldsymbol{I}]$.

The unrationalized cgs Gaussian system uses three base units: the gram, the centimetre and the second only, and derives all electric and magnetic units from appropriate combinations of them based on physical laws relating electromagnetism to mechanics. Hence, the dimensional form of electric and magnetic units in the cgs Gaussian system is $[\boldsymbol{M}]$, $[\boldsymbol{L}]$ and $[\boldsymbol{T}]$.

So the SI system for electric and magnetic units may be thought of as four dimensional and that of the unrationalized cgs Gaussian system as three dimensional.

Example: derive the dimension of the unit of charge in both the unrationalized cgs Gaussian system and that of the SI

In the unrationalized cgs Gaussian system, Coulomb's law may be written as:

$$f_G = C_G \frac{q_1 \, q_2}{r^2} \tag{10.1}$$

where f_G is the force in dynes between two electric charges q_1 and q_2, measured in esu statcoulombs or franklins, r cm apart and C_G, is a constant of proportionality. C_G is fixed, in the unrationalized case, by defining the force between two unit charges (1 Fr) separated by 1 cm to be 1 dyne, in which case $C_G = 1$. So Coulomb's law may be rewritten as:

$$f_G = \frac{q_1 \, q_2}{r^2} \tag{10.2}$$

by rearranging Equation (10.2) and taking $q_1 = q_2 = q$:

$$q^2 = f_G r^2 \tag{10.3}$$

so

$$q = f_G^{\frac{1}{2}} \, r \tag{10.4}$$

In dimensional terms, force and length may be expressed as:

$$\dim[f_G] = [M].[L].[T]^{-2} \tag{10.5}$$

$$\dim[r] = [L] \tag{10.6}$$

so the dimension of the unrationalized cgs Gaussian unit of charge $\dim[q]$ is:

$$\begin{aligned} \dim[q] &= ([L].[M].[T]^{-2})^{\frac{1}{2}} \, [L] \\ &= [M]^{\frac{1}{2}}.[L]^{\frac{3}{2}}.[T]^{-1} \end{aligned} \tag{10.7}$$

In the SI case, charge Q, in coulombs, is defined to be the product of time in seconds and the SI base unit, the ampere, so that:

$$\dim[Q] = [I].[T] \tag{10.8}$$

In Table 10.1, a selection of named electric and magnetic derived units with dimensions and combinations of base units are given for both SI and unrationalized cgs Gaussian units.

The last two entries in Table 10.1 may be defined in the following manner.

Table 10.1. *Names, symbols and dimensions of various electric and magnetic units in SI and unrationalized cgs Gaussian form*

Unit of	Name	Symbol	Dimension
Electric current SI	ampere	A	$[I]$
cgsG	esu statamp	statA	$[L]^{\frac{3}{2}}.[M]^{\frac{1}{2}}.[T]^{-2}$
Electric charge SI	coulomb	C = A . s	$[I].[T]$
cgsG	franklin	Fr	$[L]^{\frac{3}{2}}.[M]^{\frac{1}{2}}.[T]^{-1}$
Magnetic field strength SI	ampere/metre	$A.m^{-1}$	$[I].[L]^{-1}$
cgsG	oersted	Oe	$[L]^{\frac{1}{2}}.[M]^{\frac{1}{2}}.[T]^{-2}$
Electric potential difference SI	volt	V	$[L]^2.[M].[T]^{-3}.[I]^{-1}$
cgsG	esu statvolt	statV	$[L]^{\frac{1}{2}}.[M]^{\frac{1}{2}}.[T]^{-1}$
Capacitance SI	farad	F	$[L]^{-2}.[M]^{-1}.[T]^4.[I]^2$
cgsG	(esu) cm	cm	[L]
Electric resistance SI	ohm	Ω	$[L]^2.[M].[T]^{-3}.[I]^{-2}$
cgsG	esu statohm	statΩ	$[L]^{-1}.[T]$
Magnetic flux SI	weber	Wb	$[L]^2.[M].[T]^{-2}.[I]^{-1}$
cgsG	emu maxwell	Mx	$[L]^{\frac{1}{2}}.[M]^{\frac{1}{2}}$
Magnetic flux density SI	tesla	T	$[M].[T]^{-2}.[I]^{-1}$
cgsG	emu gauss	G	$[L]^{-\frac{3}{2}}.[M]^{\frac{1}{2}}$
Inductance SI	henry	H	$[L]^2.[M].[T]^{-2}.[I]^{-2}$
cgsG	esu stathenry	statH	$[L]^{-1}.[T]^2$
Electric field strength SI	volt/metre	$V.m^{-1}$	$[L].[M].[T]^{-3}.[I]^{-1}$
cgsG	esu statvolt/cm	$V.cm^{-1}$	$[L]^{-\frac{1}{2}}.[M]^{\frac{1}{2}}.[T]^{-1}$
Electric flux density SI	coulomb/sq metre	$C.m^{-2}$	$[L]^{-2}.[T].[I]$
cgsG	esu statcoulomb/sq cm	$statC.cm^{-2}$	$[L]^{\frac{3}{2}}.[M]^{\frac{1}{2}}.[T]^{-1}$
Permittivity SI of free space	farad/metre	ϵ_0	$[L]^{-3}.[M]^{-1}.[T]^4.[I]^2$
Magnetic permeability SI of free space	henry/metre	μ_0	$[L].[M].[T]^{-2}.[I]^{-2}$

The **permittivity of free space,** ϵ_0, forms part of the constant of proportionality C_{SI}, in the SI statement of Coulomb's law, i.e.:

$$f_{SI} = C_{SI}\frac{q_1 q_2}{r^2} = \frac{1}{4\pi\,\epsilon_0}\,\frac{q_1 q_2}{r^2} \tag{10.9}$$

$$C_{SI} = \frac{1}{4\pi\,\epsilon_0} \tag{10.10}$$

where f_{SI} is the force in joules between two charges q_1 and q_2, measured in coulombs, separated by a distance of r metres. The appearance of the factor 4π is due to SI units being rationalized. In general, for equations set out in rationalized units, 4π appears in those involving spherical symmetry, 2π in those with cylindrical symmetry and none in those with plane symmetry.

The **magnetic permeability of free space**, μ_0, forms part of the constant of proportionality ($\frac{\mu_0}{4\pi}$) in the equation relating the magnetic force f_M in joules, between two current elements i_1 and i_2 in amperes, in conductors 1 and 2, of length dl_1 and dl_2 metres, separated by distance $r_{1,2}$ metres, where $r_{1,2}$ is measured along a line which makes an angle θ in radians to the direction of the flow of the current in conductor 1.

$$df_M = \frac{\mu_0}{4\pi} r_{1,2}^{-2} i_1 \, dl_1 \, i_2 \, dl_2 \sin\theta \qquad (10.11)$$

Note that the constants ϵ_0 and μ_0 are related in SI units by the expression:

$$\epsilon_0 \mu_0 = \frac{1}{c^2} \qquad (10.12)$$

where c is the velocity of light (Bleaney & Bleaney, 1962).

10.2.1 A comparison between some electric and magnetic mathematical relationships in the SI and cgs Gaussian system

The various laws and definitions in the theory of electricity and magnetism take different forms depending on the units in which the variables and constants in the mathematical equations are expressed. Symbols used in Table 10.2 have the following meanings, where normal text symbols refer to scalar quantities and boldface symbols to vector quantities:

q : electric charge
i : electric current
ϵ : permittivity
ϵ_0 : permittivity of free space
μ : magnetic permeability
μ_0 : permeability of free space
c : velocity of light in vacuo
ρ : charge density
σ : conductivity

J : electric current density
E : electric intensity
B : magnetic induction
F : force
D : electric displacement
H : magnetic field intensity
V : velocity of charge element
P : polarization
M : magnetization

For a more complete treatment of the above material see, e.g., Coulson & Boyd (1979), Bleaney & Bleaney (1962) and Menzel (1960).

Table 10.2. *Some electrical and magnetic mathematical definitions in SI and unrationalized cgs Gaussian form*

Name	SI form	cgs Gaussian form
Force field	$\frac{d\mathbf{F}}{dq} = \mathbf{E} + \mathbf{V} \wedge \mathbf{B}$	$\frac{d\mathbf{F}}{dq} = \mathbf{E} + \frac{\mathbf{V} \wedge \mathbf{B}}{c}$
Electric field	$\mathbf{D} = \epsilon_0 \mathbf{E}$	$\mathbf{D} = \mathbf{E}$
	$\epsilon_0 \mu_0 = c^{-2}$	$\epsilon_0 \mu_0 = 1$
Polarization	$\mathbf{P} = \mathbf{D} - \epsilon_0 \mathbf{E}$	$\mathbf{P} = \frac{1}{4\pi}\left(\mathbf{D} - \epsilon_0 \mathbf{E}\right)$
Magnetization	$\mathbf{M} = \frac{\mathbf{B}}{\mu_0} - \mathbf{H}$	$\mathbf{M} = \frac{1}{4\pi}\left(\frac{\mathbf{B}}{\mu_0} - \mathbf{H}\right)$
Ampère's law	$\nabla \wedge \mathbf{H} = \mathbf{J} + \frac{\partial \mathbf{D}}{\partial t}$	$\nabla \wedge \mathbf{H} = \frac{4\pi \mathbf{J}}{c} + \frac{1}{c}\frac{\partial \mathbf{D}}{\partial t}$
Maxwell's equations	$\mathrm{div}\,\mathbf{D} = \rho$	$\mathrm{div}\,\mathbf{D} = 4\pi\rho$
	$\mathrm{div}\,\mathbf{B} = 0$	$\mathrm{div}\,\mathbf{B} = 0$
	$\mathrm{curl}\,\mathbf{E} = -\frac{\partial \mathbf{B}}{\partial t}$	$\mathrm{curl}\,\mathbf{E} = -\frac{1}{c}\frac{\partial \mathbf{B}}{\partial t}$
	$\mathrm{curl}\,\mathbf{H} = \sigma \mathbf{E} + \frac{\partial \mathbf{D}}{\partial t}$	$\mathrm{curl}\,\mathbf{H} = \frac{1}{c}\left(4\pi\sigma \mathbf{E} + \frac{\partial \mathbf{D}}{\partial t}\right)$
Wave equation	$\nabla^2(\mathbf{E}, \mathbf{H}) = \mu \mu_0 \epsilon \epsilon_0 \left(\frac{\partial^2}{\partial t^2}(\mathbf{E}, \mathbf{H})\right)$	$\nabla^2(\mathbf{E}, \mathbf{H}) = \frac{\epsilon \mu}{c^2}\left(\frac{\partial^2}{\partial t^2}(\mathbf{E}, \mathbf{H})\right)$

10.2.2 Converting unrationalized cgs Gaussian units to SI units

Example 1: determine the relationship between the SI unit of charge, the coulomb, and the unrationalized cgs Gaussian unit of charge, the statcoulomb
First, how many statcoulombs per charge are required to produce a force f_G of 1 N when the equal-valued charges q_G are separated, $r_G = 1$ m, apart?

The cgs Gaussian version of Coulomb's law states that:

$$f_G = \frac{q_G^2}{r_G^2} \tag{10.13}$$

substitute $r_G = 100$ cm and $f_G = 10^5$ dynes ($= 1$ newton) and rearrange Equation (10.13) to give:

$$\begin{aligned} q_G^2 &= r_G^2 f_G \\ &= 100^2 \times 10^5 = 10^9 \\ q_G &= 10^{\frac{9}{2}} \text{ statcoulombs} \end{aligned} \tag{10.14}$$

Second, how many coulombs per charge, q_{SI}, are required to produce a force f_{SI} of 1 N when the charges are $r_{SI} = 1$ m apart?

The SI version of Coulomb's law states that:

$$f_{\mathrm{SI}} = \frac{1}{4\pi\,\epsilon_0} \frac{q_{\mathrm{SI}}^2}{r_{\mathrm{SI}}^2} \tag{10.15}$$

Substitute $r_{\mathrm{SI}} = 1$ m and $f_{\mathrm{SI}} = 1$ N and rearrange Equation (10.15) to give:

$$\begin{aligned} q_{\mathrm{SI}}^2 &= \frac{1}{4\pi\,\epsilon_0} r_{\mathrm{SI}}^2 f_{\mathrm{SI}} \\ &= \frac{1}{4\pi\,\epsilon_0} \end{aligned} \tag{10.16}$$

Now ϵ_0 is a quantity defined by:

$$\epsilon_0 = \frac{1}{\mu_0 c^2} \tag{10.17}$$

where c, the speed of light in vacuo, is a defined constant, as is μ_0, the permeability of free space, hence ϵ_0 also has a fixed value. Substituting for μ_0 $(= 4\pi \times 10^{-7})$ gives:

$$\begin{aligned} \epsilon_0 &= \frac{1}{4\pi \times 10^{-7} c^2} \\ 4\pi\,\epsilon_0 &= \frac{1}{10^{-7} c^2} \\ \frac{1}{\sqrt{(4\pi\,\epsilon_0)}} &= (10^7 c^{-2})^{\frac{1}{2}} \end{aligned} \tag{10.18}$$

Comparing the values for the identical charges in the cgs Gaussian and SI units gives:

$$\begin{aligned} 10^{\frac{7}{2}} c^{-1}\,(\text{coulombs}) &\equiv 10^{\frac{9}{2}}\ \text{statcoulombs} \\ 1\,\mathrm{C} &= \frac{10^{\frac{9}{2}-\frac{7}{2}}}{c^{-1}}\,(\text{statcoulombs}) \\ &= 10c \\ &= 2.997\,924\,58 \times 10^9\ \text{statcoulombs} \end{aligned} \tag{10.19}$$

by inverting Equation (10.19):

$$1\ \text{statcoulomb} = \frac{1}{10c} = 3.335\,640\,95 \times 10^{-10}\,\mathrm{C} \tag{10.20}$$

Example 2: determine the relationship between the SI unit of magnetic flux density, the tesla (T), and the cgs Gaussian unit, the gauss (G)
First some definitions:

Magnetic flux, Φ, is usually defined in terms of magnetic flux lines as the group or number of such lines emitted outwards from the north pole of a magnetic body. Its unit is the **weber** and its symbol **Wb**.

Magnetic flux density, B, is defined to be the amount of magnetic flux per unit area perpendicular to the direction of the magnetic flux. Its unit is the **tesla** and its symbol **T**. These quantities are related by the simple expression:

$$B = \frac{\Phi}{A} \tag{10.21}$$

In SI units, when A is in m^2 and Φ is in Wb, then B is in T (teslas). In cgs Gaussian units, if A is in cm^2 and Φ is in Mx (maxwells), then B is calculated in G (gauss). The geometry of the variable in Equation (10.21) is shown in Figure 10.1.

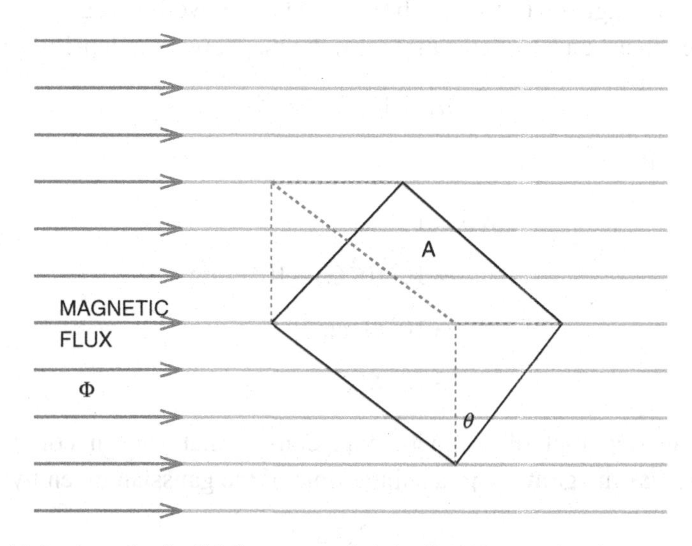

Figure 10.1. Area A of solid-line rectangle is inclined by θ rad to the perpendicular to the direction of the magnetic flux and presents an area $A\cos\theta$ to the magnetic flux and hence the magnetic flux density B, through $A\cos\theta$, is $B = \Phi / A\cos\theta$. The dotted axes (dark grey) run with and at right angles to the direction of the magnetic flux.

Next check the dimensional consistency between gauss and tesla, the dimensions for each unit are given in Table 10.1 as:

$$\dim[\text{gauss}] = [M]^{\frac{1}{2}}.[L]^{-\frac{3}{2}} \tag{10.22}$$

$$\dim[\text{tesla}] = [M].[T]^{-2}.[I]^{-1} \tag{10.23}$$

Also from Table 10.1, the cgs Gaussian dimension for electric current I_{G} is given as:

$$\dim[I]_{\mathrm{G}} = [M]^{\frac{1}{2}}.[L]^{\frac{3}{2}}.[T]^{-2} \tag{10.24}$$

substitute for $[I]$ in Equation (10.23) to give:

$$\begin{aligned}\dim[\text{tesla}] &= [M].[T]^{-2}([M]^{\frac{1}{2}}.[L]^{\frac{3}{2}}.[T]^{-2})^{-1}\\ &= [M]^{\frac{1}{2}}.[L]^{-\frac{3}{2}}\\ &= \dim[\text{gauss}]\end{aligned} \tag{10.25}$$

so the gauss and the tesla are dimensionally consistent.

Mass in cgs Gaussian units is measured in grams (g) and length in centimetres (cm); in SI units, mass is measured in kilograms (kg) and length in metres (m).

For a magnetic flux of 1 weber (Wb) passing perpendicularly through an area of 1 square metre (m^2), the magnetic flux density is $B_{\mathrm{SI}} = 1\,\mathrm{T}$ (tesla).

For a similar magnetic flux of 1 Wb ($= 10^8$ Mx (maxwells) in cgs Gaussian units) passing through a $1\,\mathrm{cm}^2$ ($= 10^{-4}\,\mathrm{m}^2$) area, the magnetic flux density is given by:

$$B_{\mathrm{G}} = 10^{-4}\,\mathrm{T} = 1\,\mathrm{G} \tag{10.26}$$

and it follows that:

$$\begin{aligned}1\,\mathrm{Wb} &= 1\,\mathrm{T}.\mathrm{m}^2\\ &= 1.(10^4\,\mathrm{G}).(10^4\,\mathrm{cm}^2)\\ &= 10^8\,\mathrm{G}.\mathrm{cm}^2\\ &= 10^8\,\mathrm{Mx}\end{aligned} \tag{10.27}$$

A cgs Gaussian unit of magnetic flux density that was in common use in astrophysics was the **gamma** γ, a submultiple of the gaussian given by:

$$1\,\gamma = 10^{-5}\,\mathrm{G} = 10^{-9}\,\mathrm{T} \tag{10.28}$$

A listing of transformations and conversions for other electric and magnetic units from cgs Gaussian to SI is given in Table 10.3.

An example of an online magnetic unit converter is given at: http://www.smpspowersupply.com/magnetic-unit-conversion.html.

Table 10.3. *Names, symbols and dimensions of various electric and magnetic units in SI and unrationalized cgs Gaussian form*

Quantity	Symbol	Transformation from cgsG to SI	cgs Gaussian unit	SI unit and conversion factor
Charge	Q	$\surd(4\pi\epsilon_0)\mathrm{Q}$	1 statcoulomb	$\frac{1}{10c}$ C
Charge density	ρ	$\surd(4\pi\epsilon_0)\rho$	1 statcoulomb . cm^{-3}	$\frac{10^5}{c}$ C . m^{-3}
Electric current	I	$\surd(4\pi\epsilon_0)\mathrm{I}$	1 statamp	$\frac{1}{10c}$ A
Current density	$\mathbf{j}$	$\surd(4\pi\epsilon_0)\mathbf{j}$	1 statamp . cm^{-2}	$\frac{10^3}{c}$ A . m^{-2}
Potential	ϕ	$\frac{\phi}{\surd(4\pi\epsilon_0)}$	1 statvolt	300 V
Electric field	$\mathbf{E}$	$\frac{\mathbf{E}}{\surd(4\pi\epsilon_0)}$	1 statvolt . cm^{-1}	3×10^4 V . m^{-1}
Magnetic flux	Φ		1 Mx	10^{-8} Wb
Magnetic flux density	$\mathbf{B}$	$\frac{\mathbf{B}}{\surd(4\pi\mu_0)}$	1 G	10^{-4} T
Vector potential	$\mathbf{A}$	$\frac{\mathbf{A}}{\surd(4\pi\mu_0)}$	1 G . cm^{-1}	10^2 c Wb . m^{-1}
Polarization	$\mathbf{P}$	$\surd(4\pi\epsilon_0)\mathbf{P}$	1 statvolt . cm^{-1}	$\frac{10^3}{c}$ C . m^{-2}
Electric displacement	$\mathbf{D}$	$\sqrt{\frac{\epsilon_0}{4\pi}}\mathbf{D}$	1 statvolt . cm^{-1}	$\frac{10^3}{4\pi c}$ C . m^{-2}
Permittivity	ϵ	$\frac{\epsilon_0}{\epsilon}$	1 statfarad . cm^{-1}	$\frac{10^7}{4\pi c^2}$ F . m^{-1}
Magnetization	$\mathbf{M}$	$\sqrt{\frac{4\pi}{\mu_0}}\mathbf{M}$	1 G	$4\pi\times10^{-4}$ A . m^{-1}
Magnetic field	$\mathbf{H}$	$\sqrt{\frac{1}{4\pi\mu_0}}\mathbf{H}$	1 Oe	$\frac{10^3}{4\pi}$ A . m^{-1}
Permeability	μ	$\frac{\mu_0}{\mu}$	1 G . Oe^{-1}	$4\pi\times10^{-7}$ H . m^{-1}
Electrical conductivity	σ	$\frac{4\pi\epsilon_0}{\sigma}$	1 s^{-1}	$\frac{10^7}{c^2}$ mho . m^{-1}

10.3 Magnetic fields in astronomy

The discovery of magnetic fields throughout space associated with stars, planets, the interplanetary and interstellar media has led to the realization of the importance of incorporating such fields and their effects in the modelling of the universe in general, and individual celestial bodies and their formation and evolution in particular. As Table 10.4 shows, the size range of magnetic fields found in the cosmos is enormous: from 10^{-10} T in interstellar space to 2.1×10^{11} T for a particular type of highly evolved star known as a magnetar. The bulk of the magnetic flux densities

Table 10.4. *Magnetic flux densities (B) of various celestial objects in cgs Gaussian and SI units*

Object	Notes	B_G G	B_{SI} T
Solar System			
Interplanetary space		1 d −6 − 1 d −5	1 d −10 − 1 d −9
Mercury	equatorial value	3.4 d −3	3.4 d −7
Venus		<4 d −6	<4 d −10
Earth		0.31	3.1 d −5
Mars		<5 d −6	<5 d −10
Jupiter		4.3	4.3 d −4
Saturn		0.22	2.2 d −5
Uranus		0.23	2.3 d −5
Neptune		1.4	1.4 d −4
Sun			
Sunspot	penumbra	0.8 d 3 − 2.0 d 3	0.08 − 0.2
	umbra	2 d 3 − 4 d 3	0.2 − 0.4
Plage		1.4 d 3 − 1.7 d 3	0.14 − 0.17
Prominences	horizontal field	2 − 40	2 d −4 − 4 d −3
Corona		1 d −5 − 1 d 2	1 d −9 − 1 d −2
Interstellar space			
	general field	1 d −6 − 1 d −5	1 d −10 − 1 d −9
Stars			
Vega	A0V star	0.6	6 d −5
FK Com	cool giant	60 − 272	6 d −3 − 2.7 d −2
Be		10 − 100	1 d −3 − 1 d −2
WR		1500	0.15
Flare		≥1000	≥0.1
T Tau		1 d 3 − 1 d 4	0.1 − 1
WD		1 d 5 − 1 d 9	10 − 1 d 5
Neutron stars		1 d 12 − 1 d 13	1 d 8 − 1 d 9
Pulsar	surface field	1 d 12	1 d 8
SGR 1806-20	magnetar	2.1 d 15	2.1 d 11
Gaseous nebulae			
Planetary		1 d −4 − 1 d −3	1 d −8 − 1 d −7
SNR		1 d −5 − 1 d −2	1 d −9 − 1 d −6
Galaxy			
$z = 0.692$		8.4 d −5	8.4 d −9

In this table the shorthand notation $m\,\mathrm{d}\,n$ has been used where $m\,\mathrm{d}\,n \equiv m \times 10^n$.

published are in the cgs Gaussian unit, the gauss, though a few are presented in the SI unit, the tesla. Both units are given in Table 10.4.

Table 10.4 was compiled, in part, from the following sources: Cox (2000), Landstreet (2001), Gnedin (1997), Reddish (1978), Bemporad & Mancusco (2010),

Table 10.5. *Solar total magnetic flux in unrationalized cgs Gaussian and SI units*

Total magnetic flux at/in	cgs Gaussian Mx	SI Wb
Solar minimum	1.5 – 2.0 d 23	1.5 – 2.0 d 15
Solar maximum	1.0 – 1.2 d 24	1.0 – 1.2 d 16
Small active region	3 d 20	3 d 12
Large active region	≥1 d 22	≥1 d 14

In this table the shorthand notation $m\,\mathrm{d}\,n$ has been used where $m\,\mathrm{d}\,n \equiv m \times 10^n$.

Table 10.6. *Planetary magnetic dipole fields in unrationalized cgs Gaussian and SI units*

Planet	cgs Gaussian $\mathrm{G\,.\,cm^{-3}}$	SI $\mathrm{T\,.\,m^{-3}}$
Mercury	2 – 6 d 22	2 – 6 d 12
Venus	<1 d 21	<1 d 11
Earth	7.84 d 25	7.84 d 15
Mars	<1 d 22	<1 d 12
Jupiter	1.55 d 30	1.55 d 20
Saturn	4.6 d 28	4.6 d 18
Uranus	3.9 d 27	3.9 d 17
Neptune	2.2 d 27	2.2 d 17

In this table the shorthand notation $m\,\mathrm{d}\,n$ has been used where $m\,\mathrm{d}\,n \equiv m \times 10^n$.

Lignières *et al.* (2009), Korhonen *et al.* (2009) and Wolfe *et al.* (2008). The surface magnetic flux density of the magnetar was extracted from the *McGill SGR/AXP Online-Catalog*.[60]

Other than magnetic flux density, Cox (2000) also lists examples of total magnetic flux for the Sun and active solar regions (see Table 10.5) and planetary magnetic dipole fields (see Table 10.6).

The relationship between cgs Gaussian units and SI units in Table 10.5 is simply: $1\,\mathrm{Wb} = 10^8\,\mathrm{Mx}$.

In Table 10.6, the relationship between the cgs Gaussian unit and the SI unit is:

$$1\,\mathrm{T\,.\,m^3} = 10^{-10}\,\mathrm{G\,.\,cm^3} \tag{10.29}$$

this follows from the relations $1\,\mathrm{G} = 10^{-4}\,\mathrm{T}$ and $1\,\mathrm{cm} = 10^{-2}\,\mathrm{m}$.

60 See http://www.physics.mcgill.ca/~pulsar/magnetar/main.html

10.3.1 Measurement of astronomical magnetic fields

Until recently, the majority of astronomical magnetic fields were detected and measured using the Zeeman effect in the observed spectrum of the celestial object. For bodies in the Solar System, including the Sun itself, magnetometers carried by various space probes have been able to measure directly their magnetic fields. At radio frequencies, techniques have been developed that depend on the way in which electrons behave in the presence of a static magnetic field to measure the strength of the field (see Burke & Graham-Smith, 2002).

Zeeman effect

The effect of a magnetic field of flux density B on the spectrum of an atom with singlet lines ($S = 0$ where S is the spin value) is to split such lines into $2J + 1$ (where J is the total angular momentum) equally spaced lines either side of the original singlet line. In cgs Gaussian units, the energies E_{M}, in ergs, of the levels are given by:

$$E_{\mathrm{M}} = E_0 \pm \frac{e\,h\,B\,M}{4\pi\,m_e\,c} \tag{10.30}$$

where E_0 is the energy level of the zero field singlet, e is the charge of the electron in statcoulombs, where:

$$1\,\mathrm{C} = 10\,c\,\mathrm{stat\,C} \tag{10.31}$$

m_e is the rest mass of the electron in grams, h is Planck's constant in erg . s, c is the speed of light in cm . s^{-1}, B is the magnetic flux density in gauss, and M takes an integer value of 0 or ± 1 (see Lang 2006).

Since:

$$E_M = h\,\nu_M \tag{10.32}$$

Equation (10.31) may be rewritten in terms of frequency ν, as (Aller, 1963):

$$\nu_M = \nu_0 \pm \frac{e\,B\,M}{4\pi\,m_e\,c} \tag{10.33}$$

where ν_M is the frequency of the singlet line split by the magnetic field and ν_0 is the frequency of the zero field singlet. Now, Equation (10.34) may be rewritten in SI units as:

$$\nu_M = \nu_0 \pm \frac{e\,B\,M}{4\pi\,m_e} \tag{10.34}$$

where ν is in hertz, e is the charge of an electron in coulombs, m_e is the rest mass of an electron in kilograms and B is the magnetic flux density in teslas.

Example: show that the dimension of the shift due to the Zeeman splitting of a singlet spectral line in both cgs Gaussian and SI units is the same

(a) cgs Gaussian case:

$$\Delta\,\nu_G = \frac{e_G\,B_G}{4\pi\,m_G\,c_G} \tag{10.35}$$

Use Table 10.1 to give the dimensions for each of the units:

$$\begin{aligned}\dim[\Delta\,\nu_G] &= \frac{\dim[e_G]\,.\dim[B_G]}{\dim[m_G]\,.\dim[c_G]} \\ &= \frac{([L]^{\frac{3}{2}}\,.\,[M]^{\frac{1}{2}}]\,.\,[T]^{-1})\,.\,([L]^{-\frac{3}{2}}\,.\,[M]^{\frac{1}{2}}\,.\,[L]\,.\,[T]^{-1})}{[L]\,.\,[M]\,.\,[T]^{-1}} \\ &= [T]^{-1}\end{aligned} \tag{10.36}$$

The unit which has a dimension of $[T]^{-1}$ is the hertz, the unit of frequency.

(b) SI case:

$$\Delta\,\nu_{SI} = \frac{e_{SI}\,B_{SI}}{4\pi\,m_{SI}} \tag{10.37}$$

Again, using Table 10.1:

$$\begin{aligned}\dim[\Delta\,\nu_{SI}] &= \frac{\dim[e_{SI}]\,.\dim[B_{SI}]}{\dim[m_{SI}]} \\ &= \frac{([I]\,.\,[T])\,.\,([M]\,.\,[T]^{-2}\,.\,[I]^{-1})}{[M]} \\ &= [T]^{-1}\end{aligned} \tag{10.38}$$

so the SI unit has a dimension of $[T]^{-1}$, showing dimensional consistency between the two systems of units.

Example: determine the numerical value for the coefficient b_{SI} in Equation (10.39)

$$b_{SI} = \frac{e_{SI}}{4\pi\,m_{SI}} \tag{10.39}$$

Substitute SI values for e_{SI} and m_{SI} in Equation (10.40):

$$\begin{aligned}b_{SI} &= \frac{1.602\,189\,072 \times 10^{-19}}{4\pi\,9.109\,3897 \times 10^{-31}} \\ &= 1.399\,634\,438 \times 10^{10}\end{aligned} \tag{10.40}$$

So the frequencies of the singlet line and the first pair of lines either side due to the presence of a magnetic field of flux density $B\,T$ are given by:

$$\nu_{SI} = \nu_0 \pm b_{SI}\,B \tag{10.41}$$

Example: at the rest frequency of the 21-cm HI line, what would be the approximate amount of Zeeman splitting resulting from a background magnetic flux density of 1 nT, similar to that found in the Milky Way Galaxy?
(a) SI case:
$\nu_{21} = 1.420\,405\,751\,768 \times 10^9$ Hz (Cox, 2000) is the radio frequency of the HI line, which corresponds approximately to a wavelength of 21.106 114 cm. The frequency separation, in a magnetic field of flux density of 10^{-9} T, is $2\,\Delta\nu_{21}$, and may be written as:

$$\begin{aligned} 2\,\Delta\nu_{SI} &= 2\,\frac{e_{SI}\,B_{SI}}{4\pi\,m_{SI}} \\ &\simeq 2\,\frac{(1.602\times10^{-19})\,(10^{-9})}{4\pi\,(9.109\times10^{-31})} \\ &\simeq 28\,\text{Hz} \end{aligned} \tag{10.42}$$

(b) cgs Gaussian case:
In cgs Gaussian units, the magnetic flux density, B_G is equal to 10^{-5} G:

$$\begin{aligned} 2\,\Delta\nu_{G} &= 2\,\frac{e_{G}\,B_{G}}{4\pi\,m_{G}\,c} \\ &\simeq 2\,\frac{(4.805\times10^{-10})\,(10^{-5})}{4\pi\,(9.109\times10^{-28})\,(2.9979\times10^{10})} \\ &\simeq 28\,\text{Hz} \end{aligned} \tag{10.43}$$

The Zeeman effect in wavelength units

The relationship between the frequency ν, and wavelength λ, of a spectral feature is given by:

$$\nu = \frac{c}{\lambda} \tag{10.44}$$

where c is the speed of light. Differentiate ν with respect to λ then:

$$\frac{d\,\nu}{d\,\lambda} = -\frac{c}{\lambda^2} \tag{10.45}$$

Replace the differentials ($d\,\nu$, $d\,\lambda$) by small differences ($\Delta\,\nu$, $\Delta\,\lambda$):

$$\Delta\,\nu = \mp\frac{c}{\lambda^2}\,\Delta\,\lambda \tag{10.46}$$

or

$$\Delta\lambda = \mp\frac{\lambda^2}{c}\Delta\nu \tag{10.47}$$

Now, in SI units, the frequency shift $\Delta\nu_{SI}$, in hertz, is given by Equation (10.38), substitute $\Delta\lambda_{SI}$ for $\Delta\nu_{SI}$ thus:

$$\begin{aligned}\Delta\lambda_{SI} &= \mp\frac{\lambda_0^2}{c}\Delta\nu_{SI}\\ &= \mp\frac{\lambda_0^2 e_{SI}}{4\pi\, m_{SI}\, c} B_{SI}\end{aligned} \tag{10.48}$$

where λ_0 is the wavelength in metres of the spectral feature in the absence of a magnetic field and $\lambda_0 \mp \Delta\lambda_{SI}$ are the central wavelengths, in metres, of the first Zeeman components formed in the presence of a magnetic field of flux density B_{SI} in teslas, with e_{SI} being the charge of the electron in coulombs and m_{SI} the mass of the electron in kilograms.

If cgs Gaussian units are used with the charge of the electron e_G measured in esu charge units (statcoulombs), the electron mass m_G in grams, the wavelengths λ in centimetres, and the speed of light in cm . s^{-1}, then by substituting for $\Delta\nu_G$ from Equation (10.36), Equation (10.48) gives:

$$\Delta\lambda_G = \mp\frac{\lambda_0^2 e_G}{4\pi\, m_G\, c^2} B_G \tag{10.49}$$

where B_G is the magnetic flux density in gauss.

In optical astronomy, the variables that are measured are λ_0 and $\Delta\lambda$, and that which is to be determined is B. Equations (10.49) and (10.50) may then be rewritten as:

$$B_{SI} = \frac{4\pi\, c\, m_{SI}\,\Delta\lambda_{SI}}{\lambda_0^2 e_{SI}} \tag{10.50}$$

and

$$B_G = \frac{4\pi\, c^2\, m_G\,\Delta\lambda_G}{\lambda_0^2 e_G} \tag{10.51}$$

Example: determine the global surface magnetic flux density of the magnetic A-type star HD94660 in both teslas and gauss

(a) SI case:

Approximate measurements were made of a small portion of the spectrum of HD94660 between 6 144 Å and 6 154 Å (as presented by J. D. Landstreet at the Leverhulme lecture on stellar magnetism.). The Zeeman doublet of interest in this narrow spectral region is due to singly ionized iron (FeII), with a laboratory-measured central wavelength of 6 149.238 Å (Moore, 1959). The separation of the minima of the Zeeman components was $2\,\Delta\lambda = 0.32\,\text{Å}$. In SI units, $\lambda_0 = 6.1492 \times 10^{-7}$ m and $\Delta\lambda = 1.60 \times 10^{-11}$ m; substitute these values into Equation (10.49) with the SI values for c, m and e:

$$
\begin{aligned}
B_{\text{SI}} &= \frac{4\pi\,.\,(9.109 \times 10^{-31})\,.\,(2.998 \times 10^{8})\,.\,(1.6 \times 10^{-11})}{(6.1492 \times 10^{-7})^2\,.\,(1.602 \times 10^{-19})} \\
&= 0.91\,\text{T}
\end{aligned}
\tag{10.52}
$$

(b) cgs Gaussian case:
In the cgs Gaussian case, wavelength is measured in cm, the speed of light in $\text{cm}\,.\,\text{s}^{-1}$, electron mass in g and electron charge in esu or statC, so Equation (10.50) becomes:

$$
\begin{aligned}
B_{\text{G}} &= \frac{4\pi\,.\,(9.109 \times 10^{-28})\,.\,(2.998 \times 10^{10})^2\,.\,(1.6 \times 10^{-9})}{(6.1492 \times 10^{-5})^2\,.\,(4.805 \times 10^{-10})} \\
&= 9100\,\text{G}
\end{aligned}
\tag{10.53}
$$

Since $1\,\text{T} = 10^4\,\text{G}$, the global magnetic field values in each set of units are equal to one another and may be compared with the value of $-2200\,\text{G}$ ($= -0.22\,\text{T}$) for the mean longitudinal magnetic field obtained by Bagnulo *et al.* (2002) using the FORS1 spectropolarimeter at the ESO VLT.

10.3.2 Magnetic fields of Solar System objects and interplanetary space from spaceborne magnetometers

The exploration of the magnetic fields of the Solar System is somewhat different from that of the rest of the universe in that it has recently become possible to send space probes to sample the fields directly rather than by using spectroscopy or spectropolarimetry. The instrument that is used to carry out measurements to determine the intensity and orientation of magnetic flux lines is known as a fluxgate magnetometer.[61]

The Mercury orbiting space probe MESSENGER carries such a magnetometer, which was able to measure the direction and magnitude of the planet's magnetic field during the January and October 2008 flybys (see Alexeev *et al.* 2010 and Anderson *et al.* 2010). Both cited papers use SI units for the equatorial surface field (given in nT) and the latter, Anderson *et al.* (2010), also presents the values for dipole and multipole fields in units of $\text{nT}\,.\,R_{\text{M}}^3$, where R_{M} is the radius of Mercury.

[61] See, e.g., http://whatis.techtarget.com/definition/

10.3.3 Faraday rotation

The plane of polarization of an electromagnetic wave is rotated in the presence of a magnetic field, with a component along the direction of propagation of the wave. This is known as the Faraday effect and the rotation of the polarization plane as Faraday rotation.

Rotation measure, R_{m}, may be defined as an integral along the line of sight from the celestial object being observed to the observer at a distance of r pc, of the product of the electron density n_e, and the line-of-sight component of the general magnetic field $B_{||}$ (the integral numerator in Equation 10.54).

Dispersion measure, D_m, is another integral along the line of sight, this time of the electron density n_e alone, from the celestial body of interest to the observer, again over a distance of r pc (the integral denominator in Equation 10.54).

Since the magnitude of the Faraday rotation is proportional to the product of the electron density and the magnetic field component along the line of sight, the ratio of the rotation measure to the dispersion measure yields the flux density of this component of the magnetic field.

Burke & Graham-Smith (2002) and Weisberg *et al.* (2004) define the magnitude of the mean line-of-sight component of the magnetic field, $< B_{||} >$, in cgs Gaussian units, along that line from the source celestial object (e.g., a pulsar), to the observer, r pc distant on Earth, to be:

$$< B_{||} > = \frac{\displaystyle\int_{s=0}^{s=r} n_e \mathbf{B} \,.\, \mathbf{ds}}{\displaystyle\int_{s=0}^{s=r} n_e \, \mathrm{d}s} \tag{10.54}$$

$$= 1.232 \times 10^{-6} \left(\frac{R_m}{D_m}\right) \mathrm{G} \tag{10.55}$$

where R_m is measured in $\mathrm{rad \,.\, m^{-2}}$ and D_m is measured in $\mathrm{pc \,.\, cm^{-3}}$. Note that whilst the dispersion measure unit may also be thought of as $\mathrm{cm^{-2}}$, the common expression used by astronomers is $\mathrm{pc \,.\, cm^{-3}}$. In SI units the dispersion measure has units of $\mathrm{m^{-2}}$ ($\equiv \mathrm{m \,.\, m^{-3}}$) or, when IAU approved units are used, the dispersion measure is measured in $\mathrm{pc \,.\, m^{-3}}$.

In cgs Gaussian units, the dimension of the quotient R_m / D_m in Equation (10.55) is given by:

$$\frac{\dim[R_m]}{\dim[D_m]} = \frac{[L] \,.\, [L]^{-1} \,.\, [L]^{-2}}{[L] \,.\, [L]^{-3}}$$

$$= 1 \tag{10.56}$$

and since $\dim[B_{||}] = [L]^{-\frac{3}{2}} . [M]^{\frac{1}{2}}$, then the dimension of the coefficient in Equation (10.55) must also be $[L]^{-\frac{3}{2}} . [M]^{\frac{1}{2}}$ to balance the dimensional equation.

In SI units, the dimensions of the quotient R_m/D_m is still unity and that of $B_{||}$ is:

$$\begin{aligned}\dim[B_{||}] &= [M].[T]^{-2}.[I]^{-1} \\ &= \frac{[M].[T]^{-2}}{[L]^{\frac{3}{2}}.[M]^{\frac{1}{2}}.[T]^{-2}} \\ &= [L]^{-\frac{3}{2}}.[M]^{\frac{1}{2}} \end{aligned} \tag{10.57}$$

The unit (rad . m^{-2}) of the rotation measure is not dependent on which of the cgs Gaussian, IAU/SI or SI units is used.

The unit of the dispersion measure in cgs Gaussian units is the pc . cm^{-3}, the appropriate unit in IAU/SI units would be pc . m^{-3} and that in pure SI units is m . m^{-3}. The conversions from cgs Gaussian to IAU/SI and SI units and gauss to teslas are:

$$1\,\text{pc.m}^{-3} = 10^6\,\text{pc.cm}^{-3} \tag{10.58}$$

$$1\,\text{m.m}^{-3} = \frac{10^6}{3.086 \times 10^{16}} = 3.240 \times 10^{11}\,\text{pc.cm}^{-3} \tag{10.59}$$

$$1\,\text{T} = 10^4\,\text{G} \tag{10.60}$$

Applying these conversion factors to the cgs Gaussian dispersion measures alters the proportionality coefficient, P_G, in the following ways:

$$P_G = 1.232 \times 10^{-6} \tag{10.61}$$

$$P_{IAU} = 1.232 \times 10^{-6} \left(\frac{10^6}{10^4}\right) = 1.232 \times 10^{-4} \tag{10.62}$$

$$P_{SI} = 1.232 \times 10^{-6} \left(\frac{10^6}{10^4 \times 3.086 \times 10^{16}}\right) = 3.992 \times 10^{-21} \tag{10.63}$$

where P_{IAU} is the proportionality coefficient for the IAU/SI unit and P_{SI} is that for the pure SI unit.

Example: evaluate the line-of-sight component of the galactic magnetic field in the direction of the pulsar J2022+2854 located near the Cygnus/Vulpecula border

Weisberg *et al.* (2004) carried out a comprehensive series of observations of Faraday rotation measures of pulsars using the Arecibo radio telescope at a frequency of

430 MHz. Included in their programme was the pulsar J2022+2854, at a distance of around 3 kpc (Verbiest *et al.*, 2010). The observed values for the rotational measure and dispersion measure were: $R_m = -73.7\,\text{rad}\,.\,\text{m}^{-2}$ and $D_m = 24.623\,\text{pc}\,.\,\text{cm}^{-3}$. Given these measurements, the mean line-of-sight magnetic flux densities, $< B_{||} >$, were calculated in each of the systems of units.

(a) cgs Gaussian units:

$$< B_{||} >_{\text{G}} = P_{\text{G}} \left(\frac{R_m}{D_m} \right) \tag{10.64}$$

where

$$R_m = -73.7\,\text{rad}\,.\,\text{m}^{-2} \tag{10.65}$$

$$D_m = 24.623\,\text{pc}\,.\,\text{cm}^{-3} \tag{10.66}$$

so

$$< B_{||} >_{\text{G}} = 1.232 \times 10^{\;6} \left(\frac{-73.7}{24.623} \right) = -3.688 \times 10^{\;6}\,\text{G} = -3.688\,\mu\,\text{G} \tag{10.67}$$

(b) IAU/SI units:

$$R_m = -73.7\,\text{rad}\,.\,\text{m}^{-2} \tag{10.68}$$

$$D_m = 2.4623 \times 10^7\,\text{pc}\,.\,\text{m}^{-3} \tag{10.69}$$

so

$$\begin{aligned} < B_{||} >_{\text{IAU}} &= 1.232 \times 10^{-4} \left(\frac{-73.7}{2.4623 \times 10^7} \right) \\ &= -3.688 \times 10^{-10}\,\text{T} = -368.8\,\text{pT} = -3.688\,\mu\text{G} \end{aligned} \tag{10.70}$$

(c) SI units:

$$R_m = -73.7\,\text{rad}\,.\,\text{m}^{-2} \tag{10.71}$$

$$D_m = \frac{2.4623 \times 10^7}{3.086 \times 10^{16}} = 7.979 \times 10^{-10}\,\text{m}\,.\,\text{m}^{-3} \tag{10.72}$$

so

$$\begin{aligned} < B_{||} >_{\text{SI}} &= 3.992 \times 10^{-21} \left(\frac{-73.7}{7.979 \times 10^{-10}} \right) \\ &= -3.687 \times 10^{-10}\,\text{T} = -368.7\,\text{pT} = -3.687\,\mu\text{G} \end{aligned} \tag{10.73}$$

So, to calculate the galactic field magnetic flux density in teslas from Faraday rotation measures, input R_m in units of rad . m^{-2}, D_m in units of m . m^{-3} and use P_{SI} as the value for the coefficient of proportionality.

10.4 Electric fields in astronomy

The presence of electric fields, potentials and currents has been detected in interplanetary space, on and about the Sun, in planetary atmospheres (e.g., the Earth, Jupiter and Saturn) and in space near planetary bodies.

10.4.1 The Stark effect on stellar spectra

The Stark effect due to an electric field is analogous to the Zeeman effect due to a magnetic field. The electric field causes the splitting of spectral lines in a manner that depends on the strength of that field. To be observable in stellar spectral features would require a uniform electric field over the entire stellar atmosphere – a phenomenon that is neither expected nor observed. However, theoretical calculations of Stark broadening parameters of likely interest to astrophysicists have recently been made for use with spectra obtained with the Far Ultraviolet Spectroscopic Explorer satellite (FUSE) by Alonso-Medina *et al.* (2009).

10.4.2 Measured electric fields, potentials and currents in, around, near and on astronomical bodies

Some examples of measurements made, and of units used, in astronomy of electric fields, potential, power and currents follow.

Solar electric fields and currents

Spectra and magnetograms of the NOAA active region 6233 were obtained at the Mees Solar Observatory by de la Beaujardière *et al.* (1993) on 28 and 29 August 1990. On the later date they found values of the maximum J_{max}, and minimum J_{min}, vertical current density to be $+26\,mA\,.\,m^{-2}$ and $-20\,mA\,.\,m^{-2}$ and the total positive current I_+, and total negative current I_-, over the entire active region to be $+8.8 \times 10^{12}$ A and -8.9×10^{12} A, respectively. Whilst the authors used SI units for these electrical measurements, they plotted their image-plane magnetograms of the longitudinal magnetic field in cgs Gaussian units (G).

Possible near-tail electric current by Mercury

From the Mercury Messenger spacecraft flybys, Anderson *et al.* (2010) have inferred the presence of a near-tail electric current density of $100\,nA\,.\,m^{-2}$ to account for low field intensities recorded by the spacecraft near the equator of Mercury. Again SI units are used for electric current density.

Jupiter's lightning

Lightning on Jupiter was detected by the Cassini spacecraft whilst enroute to Saturn in early 2001 and the observations analyzed by Dyudina *et al.* (2004). The most powerful lightning storm recorded was measured to emit 800 MW of power in the Hα line, corresponding to some 40 GW of broadband optical power.

Saturn's aurorae

A study of the electric currents in the polar ionosphere of Saturn was carried out by Cowley *et al.* (2004). They reported potentials in kV units, electron energy fluxes in mW . m^{-2}, upward field-aligned current intensities in mA . m^{-1} and particle number densities in cm^{-3} (cgs Gaussian units); in SI, the unit would be particles . m^{-3}.

10.5 Summary and recommendations

10.5.1 Summary

It is arguable that the relationships between the SI electrical and magnetic units and those based on the older cgs system are more complex than those of the other fundamental units. There are five major systems of units currently in use, designed variously by physicists, chemists, electrical engineers and metrologists. Since astronomers have, in the past, tended to favour the cgs Gaussian system of units, this chapter has concentrated mainly on the relationship between that system and the SI.

Examples are given of the names, symbols and dimensions of a range of derived electric and magnetic units in SI and unrationalized cgs Gaussian form. Of importance is the difference between the defining mathematical equations for the two unit systems, examples of which are set out in tabular form.

Worked examples of how to convert unrationalized cgs Gaussian units to SI units are given and a table of coefficients to enable ready transformations between the systems. Examples of celestial magnetic fields and units used to express them and methods of measuring them (e.g., Zeeman splitting, Faraday rotation) are described.

The chapter finishes with a brief outline of electric fields in astronomy and examples of the measurements made and units used.

10.5.2 Recommendations

Units of electricity and magnetism are manifold, and great care needs to be taken in ascertaining to which of the five major systems the unit belongs (esu, emu, cgs Gaussian, practical or SI). The conversion coefficients are not just powers of ten, but often involve the physical and mathematical constants c, μ_0, ϵ_0 and π.

Given that the IAU strongly recommends using SI units in all cases and that the most common electric or magnetic unit used by astronomers is that for magnetic flux density, such an outcome should be easy to accomplish. Perusal of relevant astronomical literature to date indicates that SI units are becoming the unit of choice in this area, so perhaps all that is needed is to encourage those still using cgs Gaussian for research or teaching to change to SI. Until such a desirable situation is achieved, a very strong recommendation is made to all astronomers when publishing results to specify precisely which system or systems of units they are using.

11
Unit of amount of substance (mole)

11.1 SI definition of the mole

1. The **mole** is the amount of substance of a system that contains as many elementary entities as there are atoms in 0.012 kilogram of carbon 12 (^{12}C); its symbol is **mol**. 2. When the mole is used, the elementary entities must be specified and may be atoms, molecules, ions, electrons, other particles, or specified groups of such particles.

The dimension of amount of substance is $[N]$.

11.1.1 Possible future definition of the mole

Presently under discussion is redefining the unit of amount of substance in the following way:

The mole is an amount of substance such that the Avogadro constant is exactly $6.022\,141\,5 \times 10^{23}$ mol^{-1} (per mole).

11.2 Avogadro's constant and atomic masses

In keeping with the proposed new definition of the mole, Avogadro's constant may be defined as the number N_A of elementary entities per mole of substance which has the (current) value $6.022\,141\,79 \times 10^{23}$ mol^{-1} (Mohr *et al.*, 2007). So the number of atoms in 0.012 kg of ^{12}C is $6.022\,141\,79 \times 10^{23}$.

Note that the dimension of Avogadro's constant is $[N]^{-1}$ and its symbol $\mathbf{mol}^{-1}$.

11.2.1 Atomic and molar masses

In SI units, the **atomic mass unit** (amu) is defined to be exactly 1/12 the mass of one atom of the ^{12}C isotope.

In SI units, the **molar mass** of a substance is defined to be the mass in kg of 1 mol of the elementary entities (e.g., atoms or molecules) composing the substance. Hence

$$0.012\,\mathrm{kg}\ {}^{12}\mathrm{C} = 1\,\mathrm{mol}\,({}^{12}\mathrm{C\ atoms}) \tag{11.1}$$

It follows that the mass $m(^{12}\mathrm{C})$ of one atom of ^{12}C in kg is:

$$\begin{aligned} \mathrm{m}(^{12}\mathrm{C}) &= \frac{0.012}{N_\mathrm{A}} \\ &= 1.992\,646\,54 \times 10^{-26}\,\mathrm{kg} \end{aligned} \tag{11.2}$$

and that the mass of one atomic mass unit is given by:

$$\begin{aligned} 1\,\mathrm{amu} &= \frac{1.992\,646\,54 \times 10^{-26}}{12} \\ &= 1.660\,538\,78 \times 10^{-27}\,\mathrm{kg} \end{aligned} \tag{11.3}$$

In cgs units (still in common use in modern university textbooks, e.g., Chang 2005) the mass of one atom of ^{12}C is $1.992\,646\,54 \times 10^{-23}$ g and that of 1 amu is $1.660\,538\,78 \times 10^{-24}$ g. The conversion factor between cgs units and SI units is 0.001 (1 g = 0.001 kg; 1 kg = 1000 g).

11.2.2 Average atomic mass

The published atomic mass of an atom in general refers to a mean value determined for a sample of that element composed of all its naturally occurring isotopes. For example, hydrogen has three isotopes, ^{1}H or hydrogen with a nucleus consisting of a single proton, ^{2}H or deuterium with a nucleus containing a single proton and a single neutron, and ^{3}H or tritium with a nucleus containing one proton and two neutrons.

Hence, the average atomic mass of hydrogen is the sum of the atomic masses of hydrogen, deuterium and tritium, each multiplied by the mole fraction of that isotope which occurs in the sample of hydrogen gas whose average atomic mass is to be determined.

The **mole fraction** is a dimensionless quantity X_i which, in the case of a sample of isotopes of a given element, expresses the ratio of the number of moles of one isotope n_i to the number of moles of all the isotopes present, n_T (see, e.g., Chang 2005).

$$X_i = \frac{n_i}{n_\mathrm{T}} \tag{11.4}$$

An example of an astronomical use of mole fractions is given by Flasar *et al.* (2005), who used early Cassini space probe infrared observations of the atmosphere of Saturn's largest moon, Titan, to determine the stratospheric mole fractions of the molecules CH_4 (methane) and CO (carbon monoxide) to be $1.6 \pm 0.5 \times 10^{-2}$ and $4.5 \pm 1.5 \times 10^{-5}$, respectively.

Example: determine the average atomic mass of carbon

In the case of carbon there are three naturally occurring isotopes: ^{12}C, ^{13}C and ^{14}C. Carbon 14 is a radioactive isotope with a half life of 5700 y, a value that has proved to be extremely useful in dating human artifacts. ^{12}C has a mass of 12 amu and is the major constituent of a typical terrestrial sample of carbon with a percentage mole fraction of 98.93 %, ^{13}C has a mass of 13.003 35 amu and a percentage mole fraction of 1.07 % and ^{14}C has a mass of 14.003 24 amu and a percentage mole fraction of $<10^{-10}$ %. The average atomic mass, $\overline{m}(C)$ of a terrestrial sample of carbon is thus:

$$\begin{aligned}\overline{m}(C) &= \frac{98.93}{100} \times 12 + \frac{1.07}{100} \times 13.003\,35 + \frac{10^{-10}}{100} \times 14.003\,24 \\ &= 12.010\,74\,\text{amu} \qquad (11.5)\end{aligned}$$

A comprehensive listing of the average atomic masses for each of the elements is given by Wieser & Berglund (2009) and a periodic table based on the data by G. P. Moss is available on the web.[62]

The average atomic mass $\overline{m}$ of an element with n isotopes, where the mole fraction of the ith isotope of mass m_i is x_i, is given by:

$$\overline{m} = \sum_{i=1}^{n} x_i\, m_i \qquad (11.6)$$

11.2.3 Molecular mass

The molecular mass of a molecule is defined to be the sum of the individual average atomic masses of the atoms which make up the molecule.

Example: calculate the molecular mass of methane (CH_4) and carbon monoxide (CO) in atomic mass units

CH_4

$$\begin{aligned}\text{molecular mass of } CH_4 &= (\text{atomic mass of C}) + 4\,(\text{atomic mass of H}) \\ &= 12.011 + 4\,(1.0079) \\ &= 16.0426\,\text{amu} \qquad (11.7)\end{aligned}$$

[62] See http://www.chem.qmul.ac.uk/iupac/AtWt/table.html

CO

$$\begin{aligned} \text{molecular mass of CO} &= (\text{atomic mass of C}) + (\text{atomic mass of O}) \\ &= 12.011 + 15.999 \\ &= 28.010\,\text{amu} \end{aligned} \tag{11.8}$$

Note that the mass of 1 mole of CH_4 is 0.016 043 kg and that of CO is 0.028 010 kg, since the molar mass in kg of a molecule is numerically equal to 0.001 times its molecular mass (there is a direct equivalence if cgs units are used, i.e., 1 mole of CH_4 has a mass of 16.043 g and 1 mole of CO has a mass of 28.010 g). For further examples of the use of the mole, see Mills *et al.* (1993).

11.2.4 Mass spectrometer

An important instrument in determining the chemical composition of astronomical objects such as meteorites, lunar and martian rock samples and the solar wind is the mass spectrometer. In the first two cases, the sample to be examined may be analyzed on Earth whilst, generally, the second pair require a miniaturized mass spectrometer flown onboard a spacecraft.

The mass spectrometer is designed to determine the ionic, atomic and molecular masses present in a gaseous sample of the object of interest. The gaseous sample is bombarded by high-energy electrons to produce positive ions that may readily be accelerated between pairs of oppositely charged plates, following which, the ions are injected into a circular path by a magnet. The radius of the circular path is a function of the ratio of e/m, with a larger value of this ratio producing a tighter radius curve. So the mass of the ion, and hence that of the atom or molecule from which it is produced, is also a function of the curve radius. Thus, it is possible to determine the ratio of the quantities of different isotopes present in a sample and the atomic constituents of a molecule.

In the case of the direct measurement of the solar wind, the detected particles are already ionized. Weygand *et al.* (1999) determined the argon and neon isotopic ratios of the solar wind from measurements made with the CELIAS isochronous time-of-flight mass spectrometer (Hovestadt *et al.*, 1995) flown onboard the SOHO spacecraft. The isotopic ratios they measured for the solar wind were:

$$\frac{^{36}\mathrm{Ar}}{^{38}\mathrm{Ar}} = 5.8 \pm 1.1 \qquad \text{and} \qquad \frac{^{20}\mathrm{Ne}}{^{22}\mathrm{Ne}} = 14.7 \pm 3.0$$

The SOHO[63] spacecraft also records the speed of the solar wind and its particle density at a given time. For example, at 23 h 44 m on 12 October 2010 the solar

[63] See http://sohowww.nascom.nasa.gov/

wind speed was 363 km . s^{-1} and its particle density in cgs units, 3.89 cm^{-3}, which is equivalent to 3.89×10^6 m^{-3} in SI units since 1 $cm^{-3} \equiv 10^6$ m^{-3}.

An example of a ground-based determination of isotopic ratios is the examination of five silicon carbide grains by Amari *et al.* (1992) obtained from the Merchison carbonaceous meteorite. The sample produced values for the $^{26}Al/^{27}Al$ isotopic ratio that varied from 0.20 ± 0.01 to 0.61 ± 0.04.

11.2.5 Some further constants involving the mole

Faraday constant

If Q is the quantity of charge measured in coulombs flowing through an electrolytic solution and n the number of moles of monovalent ions released at an electrode in that solution, then the Faraday constant F is defined by:

$$F = \frac{Q}{n} \tag{11.9}$$

The dimension of the Faraday constant is $[\boldsymbol{I}] . [\boldsymbol{T}] . [\boldsymbol{N}]^{-1}$ and its unit $\mathbf{C . mol^{-1}}$.

Since the charge carried by a single monovalent ion is equivalent to the electric charge e of an electron, then Equation (11.9) may be rewritten as:

$$F = N_{\mathrm{A}}\, e \tag{11.10}$$

where N_{A} is Avogadro's constant.

Using the values for e and N_{A} given by Mohr *et al.* (2007), the Faraday constant is equal to:

$$\begin{aligned} F &= 6.022\,141\,79 \times 10^{23} \times 1.602\,176\,487 \times 10^{-19} \\ &= 9.648\,533\,98 \times 10^{4}\,\mathrm{C.mol^{-1}} \end{aligned} \tag{11.11}$$

Universal or molar gas constant

(1) Boyle's law: at constant n (number of moles of molecules present) and T (temperature of the gas in kelvin), the volume V of the gas in m^3 is inversely proportional to the pressure of the gas in pascals ($V \propto P^{-1}$).
(2) Charles' law: at constant n and P, the volume of the gas is directly proportional to the temperature in kelvin ($V \propto T$).
(3) Avogadro's law: at constant P and T, the volume of the gas is directly proportional to the number of moles of molecules present ($V \propto n$).

These three gas laws may be combined to create a universal, ideal or perfect gas law in which:

$$R = \frac{V\,P}{n\,T} \tag{11.12}$$

where R is the universal or molar gas constant that has dimension $[L]^2 . [M] . [T] . [N]^{-1} . [\Theta]^{-1}$ and unit $\mathbf{J . mol^{-1} . K^{-1}}$, and is equal to (Mohr *et al.*, 2007):

$$R = 8.314\,472\ \mathrm{J . mol^{-1} . K^{-1}} \tag{11.13}$$

In cgs units, $R = 8.314\,472 \times 10^7\ \mathrm{erg . mol^{-1} . K^{-1}}$ (see, e.g., page 94 *et seq.* in Aller (1963) for a complete cgs treatment of the ideal gas laws as relevant to astrophysics).

Molar volume of an ideal gas

The molar volume V_m of an ideal gas is given by:

$$V_\mathrm{m} = \frac{R\,T}{P} \tag{11.14}$$

The dimension of the molar volume of an ideal gas is $[L]^3 . [N]^{-1}$ and its unit is $\mathbf{m^3 . mol^{-1}}$.

If $T = 273.15$ K and $P = 100$ kPa, then $V_\mathrm{m} = 22.710\,981 \times 10^{-3}\ \mathrm{m^3 . mol^{-1}}$. For the standard atmosphere, $T = 273.15$ K and $P = 101.325$ kPa, and the molar volume is $V_\mathrm{m} = 22.413\,996 \times 10^3\ \mathrm{m^3 . mol^{-1}}$. These values of V_m are listed in Mohr *et al.* (2007).

Molar masses of selected and subatomic particles

(1) Electron molar mass $= 5.485\,799\,0943 \times 10^{-7}\ \mathrm{kg . mol^{-1}}$
(2) Proton molar mass $= 1.007\,276\,466\,77 \times 10^{-3}\ \mathrm{kg . mol^{-1}}$
(3) Neutron molar mass $= 1.008\,664\,915\,97 \times 10^{-3}\ \mathrm{kg . mol^{-1}}$

The dimension of each of the above subatomic molar masses is $[M] . [N]^{-1}$ and its unit is $\mathbf{kg . mol^{-1}}$. These molar masses are given in CODATA recommended values of the fundamental physical constants (Mohr *et al.*, 2007).

11.3 Astrochemistry and cosmochemistry

Both astrochemistry and cosmochemistry are specialized branches of astronomy. According to Cowley (1995), cosmochemistry deals with the chemical processes responsible for the observed abundances of ions, atoms and molecules in the universe and the nuclear processes that cause such ions, atoms and molecules to form.

The short definition of astrochemistry in *The Oxford Encyclopedic English Dictionary* (Hawkins & Allen, 1991) and the rather longer one in the *Oxford Dictionary of Astronomy* (Ridpath, 2007) both define astrochemistry as a subset of cosmochemistry that deals primarily with interstellar chemistry (i.e., low temperature and pressure ambient physical conditions).

11.3.1 Interstellar chemistry

More than 100 interstellar and circumstellar molecules have been discovered (Herbst, 2001) using radio and infrared astronomy over the past 50 years, from the simplest H_2 hydrogen molecule to those comprising 10 or more atoms (e.g., $HC_{11}N$; CH_3COCH_3). The majority of these molecules are formed in interstellar clouds of low density and at low temperatures. In astrophysics, the chemical reactions which are of most importance are of the bimolecular type (Lang, 2006), e.g.,

$$A + BC \rightarrow AB + C \qquad (11.15)$$

where A and C are atoms and AB and BC are molecules. The rates of change with time of the number densities, N_A etc., are related in the following way:

$$-\frac{dN_A}{dt} = -\frac{d}{N_{BC}}dt = \frac{dN_{AB}}{dt} = \frac{dN_C}{dt} = kn_A N_{BC} \qquad (11.16)$$

Viala (1986) published a steady-state model of the chemical composition of interstellar clouds in which the rates of gas-phase reactions k are written in the general form:

$$k = (aT + b)T^{-\alpha} e^{-\frac{\beta}{T}} \text{cm}^3.\text{s}^{-1} \qquad (11.17)$$

where the coefficients a, b, α and β for 1074 reactions are given. The units of a and b are $\text{cm}^3.\text{s}^{-1}$, β is in kelvins and α is dimensionless (a ratio).

Example: derive the equation of gas-phase reactions for the chemical reaction $C + H_2O \rightarrow CO + H_2$

From Table A in Viala (1986), the values of the appropriate coefficients for the chemical reaction are:

$$a = 0\,\text{cm}^3.\text{s}^{-1} \qquad b = 2.10 \times 10^{-14}\,\text{cm}^3.\text{s}^{-1}$$

$$\alpha = -0.50 \qquad \beta = 0 \qquad (11.18)$$

Substitute these values into Equation (11.17):

$$k = 2.10 \times 10^{-14}\,T^{0.5}\,\text{cm}^3.\text{s}^{-1} \qquad (11.19)$$

In SI, the coefficients would be in units of $m^3 . s^{-1}$ (where $1\,m^3 . s^{-1} \equiv 10^6\,cm^3 . s^{-1}$) so that for the $C + H_2O \rightarrow CO + H_2$ reaction:

$$a = 0\,m^3 . s^{-1} \qquad b = 2.10 \times 10^{-8}\,m^3 . s^{-1}$$

$$\alpha = -0.50 \qquad \beta = 0 \tag{11.20}$$

$$k = 2.10 \times 10^{-8}\,T^{0.5}\,m^3 . s^{-1} \tag{11.21}$$

Table A in Viala (1986) also contains an entry, dE, for variation in enthalpy (defined by Brimblecombe *et al.* 1998 as *the heat energy associated with a chemical change*) during the chemical reaction, with a negative value indicative of an exothermic reaction and positive for an endothermic reaction. The unit used is the kcal . mol^{-1}.

Example: convert the value of the variation in enthalpy dE from cgs units to SI units for the chemical reaction $C + H_2O \rightarrow CO + H_2$
From Viala (1986), the value of dE for this reaction given in Table A is -140.076 kcal . mol^{-1} and there is an exact relationship between calories and joules, such that 1 cal $\equiv$ 4.184 J and 1 kcal $\equiv$ 4184 J, hence:

$$-140.076\,\text{kcal} . \text{mol}^{-1} = -140.076 \times 4184\,\text{J} . \text{mol}^{-1}$$

$$= -5.86078 \times 10^5\,\text{J} . \text{mol}^{-1} \tag{11.22}$$

11.4 Summary and recommendations

11.4.1 Summary

The mole, the unit of amount of substance, is almost certainly the least used of all the SI base units by astronomers as it sits more comfortably in the realm of the chemical rather than the physical sciences. The fundamental relationships between the mole and Avogadro's constant is described. This will prove to be of great importance should the SI definition of the mole be directly linked to Avogadro's constant.

Atomic, molar and molecular masses are defined, as is the atomic mass unit and the mole fraction. Examples are given of deriving the average atomic mass of carbon isotopes and the molecular masses of methane and carbon monoxide. A brief description of the mass spectrometer is followed by examples of observations made with such an instrument in the laboratory and from space.

Compound units involving the mole are described and examples given, including transforming cgs to SI units. In the final section of the chapter, astrochemistry, cosmochemistry and interstellar chemistry are defined and a worked example given of deriving the equation of gas-phase reactions for the case of $C + H_2O$.

11.4.2 Recommendations

Wilkins (1989) makes no specific reference to the mole or its usage, though the SI version is implied from the overriding recommendation from the IAU that all SI units are acceptable.

The limited appearances of the mole in modern astronomical literature means that it has to be assumed that a definition of the unit in a particular source is based on ^{12}C rather than the earlier usage of ^{16}O or ^{1}H. When compound units are used, there is a tendency for some parts of the unit to come from the cgs range (e.g., $cm^3 . mol^{-1}$ rather than $m^3 . mol^{-1}$) and, as is not uncommon in astronomical papers, textbooks or reference works, sometimes a mixture of both cgs and SI units. This practice should be avoided as confusion is the likely outcome.

12
Astronomical taxonomy

12.1 Definition of taxonomy

Taxonomy is defined as the science of classification and is derived from the ancient Greek word, ταξις, meaning arrangement, order, regularity (Liddell & Scott, 1996).

12.2 Classification in astronomy

Funk & Wagnalls New Standard Dictionary of the English Language (1946) defines classification as:

> The act or process of arranging by classes; a grouping into classes; the putting together of like objects or facts under a common designation; a process based on similarities of nature, attributes, or relations. Classification may proceed by the gathering together of similar things into a class, or by the unfolding of general groups into narrower or more specific divisions.

To classify, therefore, is to arrange in a class or classes on the basis of observed resemblances and differences. For this to proceed, two pieces of information are needed: an identity (the name of that which is to be classified) and an attribute (does the identified object have or could it have the necessary information for it to be classified as having the attribute), e.g., does Sirius (the identity – the name of the star) have a spectral type (the attribute or classification)? The answer is yes and the spectral type of Sirius is A1. It should be noted that the group of identifying names or definitions of a class of objects may also constitute a classification, e.g., the recent IAU definitions of types of bodies in the Solar System in which the new class, *dwarf planet*, contains the objects Pluto, Ceres and Eris.

For most of its history, astronomy was a purely observational science, relying entirely on receiving from the Universe electromagnetic radiation (light, infrared, ultraviolet, X-rays, γ-rays, radio waves), particular material (solar wind, cosmic rays) and larger assemblies of matter (dust, meteorites). This led to astronomy developing initially as a science of classification. Most forms of astronomical

Table 12.1. *A simple classification of the baryonic matter in the Universe*

Group	Subgroup
The Universe	Clusters of galaxies
Clusters of galaxies	Galaxies
	Intergalactic gas
	Intergalactic dust
Galaxies	Stars
	Star clusters
	Interstellar gas
	Interstellar dust
Stars	Double stars
	Exoplanets
	The Solar System
The Solar System	The Sun
	Solar wind
	The planets
	Dwarf planets
	Satellites
	Small Solar System bodies

taxonomy at the present time are quantitative, i.e., they depend on the measurement or measurements of physical attributes (position, size, distance, motion, emergent flux etc.) made using objective instrumentation. Earlier forms of taxonomy relied on a qualitative assessment of the observational attribute, e.g., in estimating the brightness of a star by visual comparison with others nearby of known brightness, or the assignment of a spectral class from the examination of a spectrogram, or determining the morphological type of a galaxy from a photographic or digital image. Whilst such subjective methods are still in use, they are far less common than they once were. Examples of objective measurements and classifications, where relevant, are given in the chapters on individual SI units. This chapter contains some examples of subjective classifications that originally depended solely on a visual estimation of some parameter or ratio, e.g., the comparison of spectral line strengths, the apparent shape of a galaxy or the magnitude of a star. Such classifications are generally dimensionless. Table 12.1 sets out one possible scheme for subdividing the Universe into classifiable groupings and subgroupings.

12.3 Classification of stellar objects

Stellar objects include stars, double stars, star clusters and planetary or gaseous nebulae.

A star itself may be defined as a spherical or ellipsoidal gaseous object that is both massive enough, and hot enough near its centre, to sustain energy-generating nuclear reactions. If the mass of the body is less than about 80 times the mass of the planet Jupiter, then the object will have insufficient mass to be able to sustain internal nuclear reactions and it would not be classified as a star.

Stars may be classified in many different ways, such as by location, brightness, colour, chemical composition, mass, age etc.

12.3.1 Stellar identity

The earliest form of identification of individual stars was to assign them names and to name prominent groupings of stars.

Individually named objects

All the bright stars and a good many fainter, naked-eye ones have individual names. For example, Sirius, the apparently brightest star in the night sky. This is the Greek name for the star, meaning sparkling or scorching (Allen, 1899). Sirius has other names in other languages. Some stars are named for their position on the sky, such as Polaris, the north polar star, or for their position within a constellation (a grouping of stars forming a generally recognizable pattern), an example of which is Betelgeuse, a bright orange–red star in the constellation of Orion, whose name loosely translates from the Arabic as 'The armpit of the giant'. Many of the proper names of stars are associated with the mythology of various races and often with their agricultural calendars or for navigational purposes.

Constellations

The entire sky is subdivided into 88 groups of stars of varying sizes, called constellations, well-known examples of which are Orion, Ursa Major (The Great Bear), Gemini (The Twins), Crux Australis (The Southern Cross) and Scorpio (The Scorpion). Some date back thousands of years and others, mainly constellations near the south celestial pole, are more recent inventions that followed the exploration of the southern hemisphere by European navigators. In the early seventeenth century, just prior to the use of the telescope for astronomical purposes, Bayer produced a star atlas (the *Uranometria*) in which the stars forming the constellations were identified with a Greek letter, assigned approximately in order of decreasing star brightness from α to ω. Thus, Sirius is also known as α Canis Majoris and Mirzam, the second brightest star in the constellation, as β Canis Majoris. If all the Greek letters are used, then the letters of the Roman alphabet or numbers assigned by Flamsteed (the first Astronomer Royal, appointed in 1675) are used.

Table 12.2. *A selection of identifiers from different catalogues for the star Sirius*

Catalogue	Identifier
Name	Sirius, Dog Star
Uranometria	α CMa, 9 CMa
Bonner Durchmusterung	BD-16 1591
5th Fundamental Catalogue	FK5 257
Henry Draper Catalogue	HD 48915
HIPPARCOS Catalogue	HIP 32349
Bright Star Catalogue	HR 2491
Infrared Astronomical Satellite	IRAS 06429-1639
ROSAT (X-ray)	RX J0645.1-1642
Smithsonian Astrophysical Observatory	SAO 151881
TD1 (Ultraviolet Photometry)	TD1 8027

Faint star identities

The advent of the telescope permitted the observation of large numbers of much fainter stars. The identity assigned to such stars is generally the name of the astronomer who constructed the star catalogue, followed by a running number. The introduction of photography resulted in another large increase in the number of stars observed, eventually resulting in the need for automated measuring machines to construct the catalogues. Stars in these publications are commonly identified by the name of the observatory or institute that carried out the photographic measurements or the name of the benefactor who provided the money for the observational programme, followed by a number that is related to the position of the star on the sky at a particular date.

By way of example, the astronomical database SIMBAD[64] lists 54 different identifiers for the star Sirius, from which the ten shown in Table 12.2 were selected.

12.3.2 Classification by brightness

The first attempt at assigning a value for the brightness of a star was due to Hipparchus, a second-century (BCE) Greek astronomer, who produced a catalogue of 1080 stars in which each star was labelled as being of the first (very bright), second, third, fourth, fifth or sixth (very faint star, only just visible to the naked eye) magnitude.

This system was extended to the 7th, 8th, 9th etc. magnitudes after the telescope revealed objects that could not be seen with the naked eye. This rather loose ordinal

[64] See http://simbad.u-strasbg.fr/simbad/

method of classifying the brightness of a star was given a little more mathematical rigour by Pogson in 1850, when he set a factor of exactly 100 as the difference in brightness between a first and a sixth magnitude star. So a one magnitude difference in brightness corresponds to a factor of $\sqrt[5]{100}$ or approximately 2.512. It should be noted that the human eye has a logarithmic response to stimulation by light.

The last major catalogue that relied on the human eye to estimate the brightness of stars was the *Bonner Durchmusterung* or BD. This contained all the stars from the north celestial pole to a declination of $-2°$ observed by Argelander with a 72 mm Fraunhofer telescope (King, 1955). The catalogue lists some 320 000 stars with positions to $0^{s}.1$ in right ascension and $0'.1$ in declination. Argelander visually estimated magnitudes to ± 0.1 to a limiting magnitude of 9.5 (van Biesbroeck, 1963). Subsequent photographically determined measures of Argelander's stars in the BD catalogue are remarkably similar to his eye estimates down to magnitude 9.0.

Visual estimates of stellar magnitudes are still made by a group of dedicated amateur astronomers whose particular interest is stars that vary in brightness over a long period of time ($\sim$100 d) or that vary erratically or unpredictably. An example of one of the techniques used is the *fractional method* (Sidgwick, 1955), where two non-varying comparison stars near to the variable star are selected, one of which is slightly brighter than the variable and the other slightly fainter (ideally the magnitude difference between the comparison stars should not exceed 0.4). The magnitude of the variable star obviously lies between the two comparison stars. The brightness interval between the comparison stars is expressed as an integer number of parts (normally less than 10) and an estimate made of the fraction of that number that best represents the brightness of the variable. If the comparison stars are of magnitudes m_1 and m_2, where $m_2 > m_1$, and the number of steps between them is n, then if the variable is estimated to be k steps brighter than the fainter comparison star, its magnitude m_{var}, will be:

$$m_{\mathrm{var}} = m_2 - \frac{k\,(m_2 - m_1)}{n} \tag{12.1}$$

12.3.3 Spectral classification of stars

The earliest serious attempt at classifying stellar spectra was carried out by Fr. Angelo Secchi. He began a spectroscopic study of the stars in 1862 (Hearnshaw, 1986), which eventually led to a classification scheme consisting of five classes. This was superseded by that of the Harvard College Observatory in the 1890s. Under the direction of E. C. Pickering, a survey of stellar spectra recorded photographically with an 11-inch telescope equipped with an objective prism were analyzed initially by Antonia Maury and later by Annie Jump Cannon. Further observations of stars in the southern hemisphere were obtained using the 13-inch

Table 12.3. *Spectral features associated with different spectral classes*

Spectral type	Effective temperature (K)	Selected spectral features
O	≥42000–34000	HeII, HeI, SiIV
B	30000–11400	HeI, SiIV, SiIII, SiII, MgII
A	9800–8100	HI, FeII, MgII, SiII, CaII, MnI
F	7300–6200	CaII, HI, FeI, CrI, CaI, CN
G	5900–5300	CaII, FeI, HI, CH, NaI
K	5200–4400	CrI, TiI, CaI, FeI, CH, CN, TiO
M	3900–3100	TiO, VO
L	2000–1300	NaI, KI, RbI, CsI, LiI, MgH, CaH, CrH, FeH MgH, CaH, CrH, FeH
T	1300–700	CH_4, H_2O, NaI, KI
(Y)	600–	

Boyden telescope in South Africa to complete the coverage of the entire sky. Classification was at first limited to magnitude 6. Since these photographs, plus others taken with a 10-inch telescope in Peru for the survey, had a much fainter limiting magnitude, the catalogue was able to be extended. Altogether, Miss Cannon classified nearly 400 000 stars by eye over a period of 45 years. The results were published in the Henry Draper catalogue plus its extensions. This gargantuan work formed the basis of the modern scheme of classifying stellar spectra.

The HD (Henry Draper) classifications are essentially one dimensional, with spectral class assigned by the relative strengths of the spectral lines of ions and atoms and the bands of molecules. The letters O, B, A, F, G, K, M, L, T, (Y) form the spectral sequence, with each class being subdivided by a number from 0 to 9, e.g., A1 (spectral type of Sirius), G2 (spectral type of the Sun) and M2 (spectral type of Betelgeuse). In physical terms, the spectral sequence from O to (Y) is one of decreasing temperature. The defining spectral features (Keenan, 1963; Kirkpatrick, 2005) and effective temperatures (Cox, 2000; Burningham *et al.*, 2008; Leggett *et al.*, 2009) of each class are given in Table 12.3. Class (Y) is not yet well defined due to a lack of suitable candidate objects.

Luminosity classification

To convert the MK classification from a one- to a two-dimensional scheme, pairs of spectral lines were selected for standard stars of known spectral type and absolute luminosity. By comparing the line ratios exhibited by stars of unknown luminosity class with those of the standard stars, an estimate of the luminosity class for the candidate star may be made. Luminosity is a function of the surface area and hence the radius of the star. If the large and small radius stars of the same spectral type have

Table 12.4. *Line pairs used to classify luminosity for different spectral class ranges (Cox, 2000)*

Spectral type range	Useful line pairs
O9 → B3	SiIV, HeI (411.6–412.2 nm) / HeI (414.4 nm)
B0 → B3	NII (399.5 nm) / HeII (400.9 nm)
B1 → A5	HI Balmer line wings
A3 → F0	MgII (441.6 nm) / MgII (448.1 nm)
F0 → F8	CaI (417.2 nm) / CaI (422.6 nm)
F2 → K5	FeI (404.5–406.3 nm) / SrII (407.7 nm)
	CaI (422.6 nm) / SrII (407.7 nm)
G5 → M	Discontinuity near 421.5 nm
K3 → M	CaI strength increasing 421.5 nm → 426.0 nm

similar masses, then the larger star must have a lower gas density and pressure, and surface gravity, than the smaller star. These differences show up in the appearance of certain spectral lines; those of a very large star are narrow and sharply defined, whilst those of a smaller radius star show the effects of pressure broadening and appear wider and less clearly defined.

Table 12.4 lists the spectral line pairs that are of particular value in assigning luminosity classes to stars of spectral type O to M. The fainter types, L, T and (Y), are for the most part not truly stars and are currently termed L dwarfs, T dwarfs and (Y) dwarfs.

There are eight main luminosity classes: class O (hypergiants), class I (supergiants), class II (bright giants), class III (giants), class IV (subgiants), class V (main sequence or dwarfs), class VI (subdwarfs), and class VII (white dwarfs). Some examples of the two-dimensional spectral type / luminosity class system are Betelgeuse (M2Iab), ϵ CMa (B2II), Arcturus (K2III), α Cru (B0.5IV), the Sun (G2V) and α Cen A (G2V).

12.3.4 Classification of star clusters

A **star cluster** may be defined as a group of stars that, forming a physical system, appear at approximately the same distance from the observer, in approximately the same direction, sharing similar transverse and radial motions.

Star clusters may be located by their appearance as an enhanced stellar density region on the sky. There are two major types of star cluster:

1. The **open** or **galactic clusters**, which typically are located in or near to the galactic plane of the Milky Way Galaxy, with member stars that generally are not centrally concentrated and have less than 1 000 members.

2. The **globular clusters**, which are strongly spherical in shape, have upwards of 1 000 members (exceeding 100 000 stars for very large examples) and are located in a roughly spherical distribution about the centre of the Milky Way Galaxy. Schemes for classifying clusters generally depend on the number of members, location relative to the galactic plane, ages, colours and spectral types of the members.

The Trumpler classification of galactic star clusters

Trumpler (1930) devised a classification scheme for galactic star clusters that uses three parameters: degree of concentration of cluster stars, range in brightness of cluster members and number of cluster members. The descriptors within each parameter are:

1. Degree of concentration

I: Detached clusters with a strong central concentration
II: Detached clusters with little central concentration
III: Detached clusters with no noticeable concentration
IV: Clusters not well detached but with strong field concentrations.

2. Range of brightness

1: Most of the cluster stars are approximately the same apparent brightness
2: A medium brightness range between the stars in the cluster
3: The cluster is composed of a mixture of bright and faint stars

3. Number of stars in the cluster

p: Poorly populated clusters with less than 50 member stars
m: Medium rich clusters with between 50 and 100 member stars
r: Rich clusters with over 100 member stars.

In addition, the letter 'n' may be added at the end of the classification to signify the presence of any form of nebulosity.

In the first instance, the degree of concentration and the range of brightness may be assessed by eye, as may the counting of those stars qualitatively presumed to be cluster members. A recent example of a Trumpler classification for the southern hemisphere open cluster IC2391 of II3r is given by Dodd (2004).

12.3.5 Morphological classification of galaxies

Even a casual glance at the illustrations of galaxies in most modern general astronomical textbooks (e.g., Kutner, 2003) reveals that galaxies do not all look the same. A scheme for classifying galaxies according to their visual appearance was devised by Hubble in the early 1920s (Hubble, 1926). A photographic selection of galaxies belonging to the different classes is set out in *The Hubble Atlas of Galaxies* by Sandage (1961) and online examples may be found on the Hubble Space Telescope website.[65]

The Hubble classification of galaxies

The majority of known galaxies may be placed on a diagram, which is generally referred to as the **tuning fork** diagram (see Figure 12.1). On the left-hand side of the diagram are found elliptical galaxies, with their degree of ellipticity, E:

$$\mathrm{E} = 10.(1 - \frac{b}{a}) \tag{12.2}$$

where a is the length of the semi-major and b that of the semi-minor axis, increasing from E0 to E7, the most elliptical type so far found. In the centre of the tuning fork are found spherically shaped galaxies with prominent bulges but no spiral arms (types S0 and SB0, where the S is for spiral and the B for barred). The upper-right portion of the diagram contains a sequence of spiral galaxies (Sa, Sb, Sc) with decreasing central bulge size relative to the entire galaxy and increasing openness of the spiral-arm structure. A parallel sequence in the lower right of the tuning fork is occupied by the barred spirals (SBa, SBb, SBc) in which the spiral arms begin, not from the galactic central bulge itself, but from the ends of a bar that runs through the galactic centre.

Some galaxies do not fit this simple classification scheme and the classes Irregular, Lenticular, Ring, Seyfert and Peculiar were added later. Table 12.5 sets out some of the more obvious morphological and astrophysical properties of the main types of galaxies.

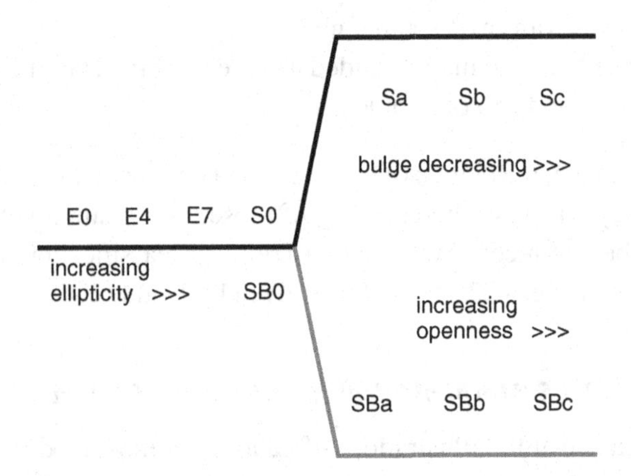

Figure 12.1. Hubble classification of galaxies.

[65] http://hubblesite.org/newscenter/archive/images/galaxy

Table 12.5. *Morphological and astrophysical properties of elliptical and spiral galaxies*

Hubble type	Nuclear bulge	Central bar	General shape	Young stars	Gas and dust	Integrated colour
E	no	no	ellipsoidal	no	small amounts	red
S	yes	no	flattened	yes	yes	blue
SB	yes	yes	flattened	yes	yes	blue

12.4 Classification of Solar System objects

In 2006, at the 26th General Assembly of the International Astronomical Union (IAU), resolutions were discussed on the definitions of the terms planet, dwarf planets and small Solar System bodies. The IAU press release was as follows:

Contemporary observations are changing our understanding of planetary systems, and it is important that our nomenclature for objects reflect our current understanding. This applies in particular, to the designation *planets*. The word planet originally described wanderers that were known only as moving lights in the sky. Recent discoveries lead us to create a new definition, which we can make using currently available scientific information.

12.4.1 Definition of a planet in the Solar System

A **planet** is a celestial body that (a) is in orbit around the Sun, (b) has sufficient mass for its self-gravity to overcome rigid-body forces so that it assumes a hydrostatic equilibrium (nearly round) shape, (c) has cleared the neighbourhood around its orbit, and (d) is not a satellite.

The eight objects defined now as planets are, in increasing distance from the Sun: Mercury, Venus, Earth, Mars, Jupiter, Saturn, Uranus, and Neptune.

12.4.2 Definition of a dwarf planet in the Solar System

A **dwarf planet** is a celestial body that (a) is in orbit around the Sun, (b) has sufficient mass for its self-gravity to overcome rigid-body forces so that it assumes a hydrostatic equilibrium (nearly round) shape, (c) has not cleared the neighbourhood around its orbit, and (d) is not a satellite.

The dwarf planet Pluto is recognized as an important prototype of a new class of trans-Neptunian objects.

Examples of dwarf planets are: Pluto, Ceres and Eris.

12.4.3 Definition of small Solar System bodies

All other bodies, except satellites, orbiting the Sun shall be referred to collectively as **small Solar System bodies**.

These currently include most of the Solar System asteroids, most of the trans-Neptunian objects (TNOs), comets, and other small bodies.

Examples of such small Solar System bodies are: Vesta, Pallas, Eros (all asteroids); Comet Halley, Comet McNaught (both comets); and meteors (whilst still in orbit about the Sun).

12.4.4 Definition of a satellite

The IAU did not produce a formal definition for a satellite. Funk *et al.* (1946) define such a body as:

> A smaller body attending and revolving round a larger one; commonly a secondary planet round a primary one.

The Earth's Moon is a satellite, as are: Phobos and Deimos of the planet Mars; Io, Ganeymede, Europa and Callisto of the planet Jupiter; Iapetus, Mimas and Titan of the planet Saturn; Umbriel, Titania and Oberon of the planet Uranus; Triton and Nereid of the planet Neptune; and Charon of the dwarf planet Pluto.

12.5 Astronomical databases and virtual observatories

Until recently, information about celestial bodies was generally published as a printed volume or volumes (e.g., *The Astronomical Almanac* (annual), the Bright Star Catalogue, the HD catalogue of stellar spectra). The advent of computer systems has led to the production of machine readable catalogues, which may be held online or made available as CD or DVD discs.

12.5.1 Astronomical databases

Both general-purpose databases and those with a particular astronomical research area in mind have been set up online. The following is a small selection of some of the data that are available, with their web address.

SIMBAD astronomical database is a compendium of information about stars, including their positions, distances, proper motions, radial velocities, photometry, spectroscopy and bibliography.
http://simbad.u-strasbg.fr/simbad/

Table 12.6. *Members of the International Virtual Observatory Alliance*

Virtual observatory	web address
International Virtual Observatory Alliance	http://www.ivoa.net/
Armenian Virtual Observatory	http://www.aras.am/Arvo/arvo.htm
Virtual Observatory United Kingdom	http://www.astrogrid.org/
Australian Virtual Observatory	http://www.aus-vo.org/
Chinese Virtual Observatory	http://www.china-vo.org/
Canadian Virtual Observatory	http://services.cadc-ccda.hia-iha.nrc-cnrc.gc.ca/cvo/
European Virtual Observatory	http://www.euro-vo.org/
German Astrophysical Virtual Observatory	http://www.g-vo.org/
Hungarian Virtual Observatory	http://hvo.elte.hu/en/
Japanese Virtual Observatory	http://jvo.nao.ac.jp/
Korean Virtual Observatory	http://kvo.kasi.re.kr/
National Virtual Observatory United States	http://www.us-vo.org
Observatoire Virtuel France	http://www.france-vo.org/
Russian Virtual Observatory	http://www.inasan.rssi.ru/eng/rvo/
Spanish Virtual Observatory	http://svo.laeff.inta.es/
Italian Virtual Observatory	http://vobs.astro.it/
Virtual Observatory India	http://vo.iucaa.ernet.in/

The European Southern Observatory (ESO) archive includes images and data, the digitized sky surveys, the HIPPARCOS and TYCHO catalogues, records of images obtained with the various ESO telescopes and ESO publications.
http://archive.eso.org/

The image and catalogue archive of the United States Naval Observatory (USNO) includes the astrometric USNO catalogues.
http://www.nofs.navy.mil/data/fchpix/

The Washington Double Star Catalogue is also housed at the USNO.
http://ad.usno.navy.mil/wds/

For astronomers with a particular interest in star clusters, the WEBDA site of the University of Geneva will prove very useful.
http://obswww.unige.ch/webda/

12.5.2 Virtual observatories

The **virtual observatory** (VO) combines astronomical databases at different centres with specially written software to create a scientific research environment that facilitates work on a variety of astronomical, astrophysical and cosmological research programmes.

There are a number of virtual observatories throughout the world. Their web addresses are given above in Table 12.6, along with that of the International Virtual Observatory Alliance (IVOA), whose primary task is to develop agreed interoperability standards with the member VOs.

12.6 Summary and recommendations

12.6.1 Summary

The classification of celestial bodies according to particular attributes is often a useful first stage in grouping together objects that may later be shown to have similar physical properties. In the examples given in this chapter, this expectation is realized with the morphological classification of galaxies, but not with the grouping of stars according to their apparent observed brightness, since stars do not all have the same intrinsic brightness and do not all lie at the same distance from the observer.

The development of astronomical databases and virtual observatories allows high-quality data to be examined by a far wider group of research astronomers than was previously the case and should lead to a greater understanding of the significance of the data.

12.6.2 Recommendations

When celestial bodies are classified by measurement (e.g., positions, distances, motions, flux outputs, temperatures etc.) and published in an online database or virtual observatory, then such measurements should be given in SI units. If other units are used, then they should be accompanied by SI units.

References

Abdo, A. A., Ackermann, M., Ajello, M. *et al.* (2009). Discovery of pulsations from the pulsar J0205+6449 in SNR 3C 58 with the FERMI γ-ray space telescope *ApJL*, **699**, L102–L107.

Abdo, A. A., Ackermann, M., Ajello, M. *et al.* (2010). The first catalog of active galactic nuclei detected by the FERMI Large Area Telescope. *ApJ*, **715**, 429–457.

Abramowitz, M. & Stegun, I. A. eds (1972). *Handbook of Mathematical Functions*. New York: Dover Publications Inc., p. 67.

Albacete-Colombo, J. F., Damiani, F., Micela, G., Sciortino, S. & Harnden, F. R., Jr. (2008). An X-ray survey of low-mass stars in Trumpler 16 with CHANDRA. *A&A*, **490**, 1055–1070.

Alexeev, I. I., Belenkaya, E. S., Slavin, J. A. *et al.* (2010). Mercury's magnetospheric magnetic field after the first two MESSENGER flybys. *Icarus*, **209**, 23–39.

Alder, K. (2004). *The Measure of All Things*. London: Abacus.

Allen, C. W. (1951). Dex. *Observatory*, **71**, 157.

Allen, R. H. (1899). *Star Names and their Meanings*. New York: G. E. Stechert & Co.

Aller, L. H. (1963). *Astrophysics: The Atmospheres of the Sun and Stars*, 2nd edn. New York: The Roland Press Company.

Alonso-Medina, A., Colón, C., Montero, J. L. & Nation, L. (2009). Stark broadening of PbIV spectral lines of astrophysical interest. *MNRAS*, **401**, 1080–1090.

Amari, S., Hoppe, P., Zinner, E. & Lewis, R. S. (1992). Interstellar SiC with unusual isotopic compositions: Grains from a supernova? *ApJ*, **394**, L43–L46.

Anderson, B. J., Acuña, M. H., Korth, H. *et al.* (2010). The magnetic field of Mercury. *Space Sci. Rev.*, **152**, 307–339.

Astronomical Almanac for 1995, The (1994). London: HMSO.

Atkin, K. (2007). Size matters. *A&G*, **48**, 4.7.

Atwood, W. B. Abdo, A. A., Ackermann, M. *et al.* (2009). The Large Area Telescope on the FERMI γ-ray space telescope mission. *ApJ*, **697**, 1071–1102.

Bagnulo, S., Szeifert, T., Wade, G. A., Landstreet, J. D. & Mathys, G. (2002). Measuring magnetic fields of early type stars with FORS1 at the VLT. *A&A*, **389**, 191–201.

Barlow, C. W. C. & Bryan, G. H. (1956). *Elementary Mathematical Astronomy*. London: University Tutorial Press Ltd., p. 125.

Battat, J. B. R., Murphy, T. W., Jr., Adelberger, E. G. *et al.* (2009). The Apache Point Lunar-ranging Operation (APOLLO): Two years of millimeter-precision measurements of the Earth–Moon range. *PASP*, **121**, 29–40.

de la Beaujardière, J.-F., Canfield, R. C. & Leka, K. D. (1993). The morphology of flare phenomena, magnetic fields and electric currents in active regions. III NOAA active region 6233 (1990 August). *ApJ*, **411**, 378–382.

Bedding, T. R., Butler, R. P., Carrier, F. *et al.* (2006). Solar-like oscillations in the metal-poor subgiant ν Indi: constraining the mass and age using asteroseismology. *ApJ*, **647**, 558–563.

Beech, M. (2008). The reluctant parsec and the overlooked light-year. *Observatory*, **128**, 489–494.

Bemporad, A. & Mancusco, S. (2010). First complete determination of plasma physical parameters across a coronal mass ejection driven shock. *ApJ*, **720**, 130–143.

Bessel, F. W. (1838). On the parallax of 61 Cygni. *MNRAS*, **4**, 152–161.

Bessell, M. S. (1992). In *The Astronomy and Astrophysics Encyclopedia*, ed. S. P. Maran. Cambridge: Cambridge University Press, p. 403.

Bessell, M. S. (2001). In *Encyclopedia of Astronomy & Astrophysics*, Vol. 2, ed. P. Murdin, Bristol: IoP, pp. 1638–1644.

BIPM (2006). *The International System of Units*, 8th edn. France: CIPM.

Blanco, V. M. & McCuskey, S. W. (1961). *Basic Physics of the Solar System*. Reading, MA: Addison-Wesley Publishing Company, Inc.

Blauuw, A., Gum, C. S., Pawsey, J. L. & Westerhout, G. (1960). The new IAU system of galactic coordinates (1958 revision). *MNRAS*, **121**, 123–131.

Bleaney, B. I. & Bleaney, B. (1962). *Electricity and Magnetism*. Oxford: Clarendon Press.

Bohlin, R. C. & Gilliland, R. L. (2004). Hubble Space Telescope absolute spectrophotometry from the far ultraviolet to the infrared. *AJ*, **127**, 3508–3515.

Bok, B. J. (1937). *The Distribution of the Stars in Space*. Chicago, IL: The University of Chicago Press.

Bradley, P. A. (1993). Methods of asteroseismology for white dwarf stars. *Baltic Astr.*, **2**, 545–558.

Brimblecombe, S., Gallannaugh, D. & Thompson, C., eds. (1998). *The Hutchinson Encyclopedia of Science*. Oxford: Helicon

Brown, M. E. & Schaller, E. L. (2007). The mass of the dwarf planet Eris. *Science*, **316**, 1585.

Browning, D. R., ed. (1969). *Spectroscopy*. London: McGraw-Hill.

Burke, B. F. & Graham-Smith, F. (2002). *An Introduction to Radio Astronomy*. 2nd edn. Cambridge: Cambridge University Press.

Burningham, B., Pinfield, D. J., Leggett, S. K. *et al.* (2008). Exploring the substellar temperature region down to ~ 500 K. *MNRAS*, **391**, 320–333.

Cairns, W. (2007). *About the Size of It*. London: MacMillan.

Capitaine, N. & Guinot, B. (2008). The astronomical units. arXiv:0812.2970v1 [astro-ph].

Cardarelli, F. (2003). *Encyclopedia of Scientific Units, Weights and Measures*. London: Springer.

Chambers, G. F. (1889). *Astronomy. I: The Sun, Planets and Comets*. Oxford: Clarendon Press.

Chang, R. (2005). *Chemistry*, 8th edn. New York: McGraw-Hill.

Cohen, M., Walker, R. G., Barlow, M. J. & Deacon, J. R. (1992). Spectral irradiance calibration in the infrared. I. Ground-based and IRAS broadband calibrations. *AJ*, **104**, 1650–1657.

Cohen, M., Wheaton, W. A. & Megeath, S. T. (2003). Spectral irradiance calibration in the infrared XIV: The absolute calibration of 2MASS. *AJ*, **126**, 1090–1096.

Collins, II, G. W. (1989). *The Fundamentals of Stellar Astrophysics*. New York: W. H. Freeman and Company.

Coulson, C. A. & Boyd, T. J. M. (1979). *Electricity*. London: Longman.
Cowley, C. R. (1995). *An Introduction to Cosmochemistry*. Cambridge: Cambridge University Press.
Cowley, S. W. H., Bunce, E. J. & Prangé, R. (2004). Saturn's polar ionospheric flows and their relation to the main auroral oval. *Ann. Geophys.*, **22**, 1379–1394.
Cox, A. N. (2000). *Allen's Astrophysical Quantities*, 4th edn, ed. A. N. Cox. New York: Springer.
Culhane, J. L. & Sanford, P. W. (1981). *X-ray Astronomy*. London: Faber & Faber.
Dodd, R. J. (2004). Data mining in the young open cluster IC2391. *MNRAS*, **355**, 959–973.
Dodd, R. J. (2007). Unified absolute spectrophotometry for star clusters, ed. C. Sterken. *ASP Conference Series*, **364**, pp. 237–254.
Duffard, R., Ortiz, J. L., Santos Sanz, P. *et al.* (2008). A study of the photometric variations on the dwarf planet (136199) Eris. *A&A*, **479**, 877–881.
Dyson, F. W. (1913). Report of the Royal Astronomical Society meeting on 14 March 1913. *Observatory*, **36**, 160.
Dyudina, U. A., del Genio, A. D., Ingersoll, A. P. *et al.* (2004). Lightning on Jupiter observed in the Hα line by the Cassini imaging science subsystem. *Icarus*, **172**, 24–36.
Emerich, C., Lamarre, J. M., Gispert, R. *et al.* (1988). Temperature of the nucleus of comet Halley. *ESA Proceedings of International Symposium on the Diversity and Similarity of Comets*, pp. 703–706.
Evans, D. S. (1954). *Teach Yourself Astronomy*. London: English Universities Press Ltd.
Evensen, K. M., Wells, J. S., Petersen, F. R. *et al.* (1972). Speed of light from direct frequency and wavelength measurements of the methane-stabilized laser. *PRL*, **29**, 1346–1349.
Ferdman, R. D., Stairs, I. H., Kramer, M. *et al.* (2010). The precise mass measurement of the intermediate-mass binary pulsar PSR J1802–2124. *ApJ*, **711**, 764–771.
Ferriere, K. (2001). The interstellar environment of our galaxy. *Rev. Mod. Phys.*, **73**, 1031–1066.
Fixler, J. B., Foster, G. T., McGuirk, J. M. & Kasevich, M. A. (2007). Atom interferometer measurement of the Newtonian constant of gravity. *Science*, **315**, 74–77.
Flasar, F. M., Achterberg, R. K., Conrath, B. J. *et al.* (2005). Titan's atmospheric temperatures, winds and composition. *Science*, **308**, 975–978.
Fouqué, P., Chevallier, L., Cohen, M. *et al.* (2000). An absolute calibration of DENIS. *A&AS*, **141**, 313–317.
Funk I. K., Thomas C., Vizetelly F. M. & Funk C. E. eds. (1946). *Funk & Wagnalls New Standard Dictionary of the English Language*, Vols I and II. London: The Waverley Book Company Ltd.
Gezari, D. Y., Schmitz, M., Pitts, P. S. & Mead, J. M. (1993). *Catalog of Infrared Observations*, 3rd edn. Greenbelt, MD: NASA reference publication, p. 1294.
Girard, G. (1994). The third periodic verification of national prototypes of the kilogram (1988–1992). *Metrologia*, **31**, 317–336.
Gnedin, Y. N. (1997). Chromospheres, activity and magnetic fields. In *Fundamental Stellar Properties: The Interaction Between Observation and Theory*, ed. T. R. Bedding *et al.* Dordrecht: Kluwer, pp. 245–252.
Gould, A., Udalski, A., Monard, B. *et al.* (2009). The extreme microlensing event OGLE-2007-BLG-224: Terrestrial parallax of a thick-disk brown dwarf. *ApJL*, **698**, 147–151.
Hansen, C. J., Kawaler, S. D. & Trimble, V. (2004). *Stellar Interiors: Physical Principles, Structure and Evolution*. New York: Springer-Verlag.

Harris, III, D. L., Strand, K. Aa. & Worley, C. E. (1963). Empirical data on stellar masses, luminosities and radii. In *Basic Astronomical Data*, ed. K. Aa. Strand. Chicago IL: University of Chicago Press, pp. 273–292.

Hawkins, J. M. & Allen, R. eds. (1991). *The Oxford Encyclopedic English Dictionary*. Oxford: Clarendon Press.

Hearnshaw, J. B. (1986). *The Analysis of Starlight*. Cambridge: Cambridge University Press.

Hearnshaw, J. B. (1996). *The Measurement of Starlight: Two Centuries of Astronomical Photometry*. Cambridge: Cambridge University Press.

Henriksen, M. J. & Tittley, E. R. (2002). CHANDRA observations of the A3266 galaxy cluster merger. *ApJ*, **577**, 701–709.

Herbst, E. (2001). In *Encyclopedia of Astronomy & Astrophysics*, Vol. 2, ed. P. Murdin, Bristol: IoP. pp. 1258–1266.

Hohle, M. M. Eisenbeiss, T., Mugrauer, M. *et al.* (2009). Photometric study of the OB star clusters NGC1502 and NGC2169 and mass estimation of their members at the University Observatory Jena. *AN*, **330**, 511.

Holland, W. S., Greaves, J. S., Zuckerman, B. *et al.* (1998). Submillimeter images of dusty debris around nearby stars. *Nature*, **392**, 788–791.

Hollis, J. M., Chin, G. & Brown, R. L. (1985). An attempt to detect mass loss from α Lyrae with the VLA. *ApJ*, **294**, 646–648.

Hovestadt, D., Hilchenbach, M., Bürgi, A. *et al.* (1995). CELIAS: Charge, Element and Isotope Analysis System for SOHO. *Sol. Phys.*, **162**, 441–481.

Huang, T.-Y., Han, C.-H., Yi, Z.-H., & Xu, B.-X. (1995). What is the astronomical unit of length? *A&A*, **298**, 629–633.

Hubble, E. P. (1926). Extragalactic nebulae. *ApJ*, **64**, 321–369.

Hubble, E. P. (1929). A relation between distance and radial velocity among extra-galactic nebulae. *PNAS*, **15**, 168–173.

Irwin, J. A. (2007). *Astrophysics: Decoding the Cosmos*. Chichester: John Wiley & Sons Ltd.

Józsa, G. I. G., Garrett, M. A., Oosterloo, T. A. *et al.* (2009). Revealing Hanny's Voorwerp: radio observations of IC2497. *A&A*, **500**, L33–L36.

Karovska, M. & Sasselov, D. (2001). In *Encyclopedia of Astronomy & Astrophysics*, vol. 4, ed. P. Murdin. Bristol: IoP, pp. 3068.

Kaye, G. W. C. & Laby, T. H. (1959). *Tables of Physical and Chemical Constants and some Mathematical Functions*, 12th edn. London: Longmans, Green and Co.

Keenan, P. C. (1963). Classification of stellar spectra. In *Basic Astronomical Data*, ed. K. Aa. Strand. Chicago: University of Chicago Press, pp. 78–122.

King, H. C. (1955). *The History of the Telescope*. London: Charles Griffin & Co. Ltd.

Kirkpatrick, J. D. (2005). New spectral types L and T. *ARA&A.*, **43**, 195–246.

Klioner, S. A. (2007). Relativistic scaling of astronomical quantities and the system of astronomical units. *A&A*, **478**, 951–958.

Korhonen, H., Hubrig, S., Berdyugina, S. V. *et al.* (2009). First measurement of the magnetic field on FK Com and its relation to the contemporaneous star-spot locations. *MNRAS*, **395**, 282–289.

Kurucz, R. L. (1979). Model atmospheres for G, F, A, B and O stars. *ApJS*, **40**, 1–340.

Kutner, M. L. (2003). *Astronomy: A Physical Perspective*, 2nd edn. Cambridge: Cambridge University Press.

Landstree, J. D. (2001). In *Encyclopedia of Astronomy & Astrophysics*, vol. 2, ed. P. Murdin. Bristol: IoP, pp. 1508–1514.

Lang, K. R. (2006). *Astrophysical Formulae*. Vols I and II: Berlin: Springer-Verlag.

Leggett, S. K., Cushing, M. C., Saumon, D. *et al.* (2009). The physical properties of four 600 K T dwarfs. *ApJ*, **695**, 1517–1526.

Leschiutta, S. (2001). In *Encyclopedia of Astronomy & Astrophysics*, vol. 4, ed. P. Murdin. Bristol: IoP, pp. 3313–3315.

Liddell, H. G. & Scott, R. (1996). *A Greek – English Lexicon*. Oxford: Clarendon Press.

Lignières, F., Petit, P., Böhm, T. & Aurière, M. (2009). First evidence of a magnetic field on Vega. *A&A*, **500**, L41–L44.

Longair, M. S. (1989). *Royal Observatory Edinburgh: Research and Facilities Handbook.* Edinburgh: ROE, p. 79.

Love, S. G. & Brownlee, D. E. (1993). A direct measurement of the terrestrial accretion rate of cosmic dust. *Science*, **262**, 550–553.

Lovell, B. & Clegg, J. A. (1952). *Radio Astronomy*. London: Chapman & Hall Ltd.

Lucas, P. W., Tinney, C. G., Burningham, B. *et al.* (2010). Discovery of a very cool brown dwarf amonst the ten nearest stars to the Solar System. arXiv:1004.0317v1[astro-ph.SR].

Lyne, A. G., Pritchard, R. S. & Graham-Smith, F. (1993). 23 years of Crab pulsar rotational history. *MNRAS*, **265**, 1003–1012.

McCarthy, D. D. & Petit, G. (2003). *IERS Technical Note, 32. IERS Conventions (2003).* Frankfurt am Main: Verlag des Bundesamtes für Kartographie und Geodäsie.

Manchester, R. N., Hobbs, G. B., Teoh, A. & Hobbs, M. (2005). The Australia Telescope National Facility pulsar catalogue. *AJ*, **129**, 1993–2006,

Mather, J. C., Cheng, E. S., Cottingham, D. A. *et al.* (1994). Measurement of the cosmic microwave background spectrum by the COBE FIRAS instrument. *ApJ*, **420**, 439–444.

Mayes, V. (1994). Unit prefixes for use in astronomy. *QJRAS*, **35**, 569–572.

Menzel, D. H. (1960). *Fundamental Formulas of Physics*. Vols 1 and 2. New York: Dover Publications Inc.

Mills, I., Cvitaš, T., Homann, K., Kallay, N. & Kuchitsu, K. (1993). *Quantities, Units and Symbols in Physical Chemistry*, 2nd edn. International Union of Pure and Applied Chemistry. Oxford: Blackwell Science.

Mills, I. M., Mohr, P. J., Quinn, T. J., Taylor, B. N. & Williams, E. R. (2005). Redefinition of the kilogram: a decision whose time has come. *Metrologia*, **42**, 71–80.

Miura, T., Arakida, H., Kasai, M. & Kuramata, S. (2009). Secular increase of the astronomical unit: a possible explanation in terms of the total angular-momentum conservation law. *PASJ*, **61**, 1247–1250.

Moffatt, J. (1950). *A New Translation of the Bible*. London: Hodder and Stoughton Ltd.

Mohr, P. J., Taylor, B. N. & Newell, D. B. (2007). The fundamental physical constants. *Phys. Today*, **60**(7), 52–55.

Monet, D. G., Levine, S. E., Canzian, B. *et al.* (2003). The USNO-B Catalog. *AJ*, **125**, 984–993.

Moore, C. E. (1959). *A Multiplet Table of Astrophysical Interest*. Washington, DC: US Department of Commerce.

Moore, P. A. (2001). In *Encyclopedia of Astronomy & Astrophysics*, vol. 1, ed. P. Murdin. Bristol: IoP, p. 464.

Muhleman, D. O., Holdridge, D. B. & Block, N. (1962). The astronomical unit determined by radar reflections from Venus. *AJ*, **67**, 191–203.

Murray, C. A. (1989). The transformation of coordinates between the systems of B1950.0 and J2000.0, and the principal galactic axes referred to J2000.0. *A&A*, **218**, 325–329.

Noerdlinger, P. D. (2008). Solar mass loss, the astronomical unit, and the scale of the Solar System. 2008arXiv0801.3807N, 1–31.

Pannekoek, A. (1961). *A History of Astronomy*. London: George Allen & Unwin Ltd.

Patilla, P. (2000). *Measuring Up Size*. London: Belitha Press Ltd.

Pease, D. O., Drake, J. J. & Kashyap, V. L. (2006). The darkest bright star: CHANDRA X-ray observations of Vega. *ApJ*, **636**, 426–431.

Penzias, A. A. & Wilson, R. W. (1965). A measurement of excess antenna temperature at 4080 Mc/s. *ApJ*, **142**, 419–421.

Perryman, M. A. C., Lindegren, L., Kovalevsky, J. *et al.* (1997). The Hipparcos Catalogue. *A&A*, **323**, L49–L52.

Pitjeva, E. V. (2005). High-precision ephemerides of planets – EPM and determination of some astronomical constants. *Solar System Res.*, **39**, 176–186.

Pitjeva, E. V. & Standish, E. M. (2009). Proposals for the masses of the three largest asteroids, the Earth – Moon mass ratio and the astronomical unit, *Celest. Mech. Dyn. Astr.*, **103**, 365–372.

Reddish, V. C. (1978). *Stellar Formation*. Oxford: Pergamon Press.

Ridpath, I. (2007). *Oxford Dictionary of Astronomy*, 2nd edn. Oxford: Oxford University Press.

Rufener, F. & Nicolet, B. (1988). A new determination of the Geneva photometric passbands and their absolute calibration. *A&A*, **206**, 357–374.

Rutherford, E. (1929). Origin of actinium and age of the Earth. *Nature*, **123**, 313–314.

Sackman, I.-J., Boothroyd, A. I. & Kraemer, K. E. (1993). Our Sun. III: Present and future. *ApJ*, **418**, 457–468.

Saha, M. N. (1921). On the physical theory of stellar spectra. *Proc. Roy. Soc. London*, **A99**, 135–138.

Sandage, A. (1961). *The Hubble Atlas of Galaxies*. Publications of the Carnegie Institution of Washington No. 618, Washington.

Schramm, D. N. (1990). The Age of the Universe: Concordance. In *Astrophysical Ages and Dating Methods*, ed. E. Vangioni–Flam, *et al.* Gif-sur-Yvette, France: Editions Frontières, pp. 365–383.

Shane, C. D. & Wirtanen, C. A. (1954). The distribution of the extragalactic nebulae. *AJ*, **59**, 285–306.

Shklovskii, I. S. (1970). Possible causes of the scalar increase in pulsar periods. *Sov. Astr.*, **13**, 562–565.

Sidgwick, J. B. (1955). *Observational Astronomy for Amateurs*. London: Faber & Faber Ltd.

Spencer-Jones, H. (1956). *General Astronomy*. London: Edward Arnold (Publishers) Ltd.

Stencel, R. E. Creech-Eakman, M., Hart, A. *et al.* (2008). Interferometric studies of the extreme binary ϵ Aurigae: pre-eclipse observations. *ApJL*, **689**, L137–L140.

Sterken, C. & Manfroid, J. (1992). *Astronomical Photometry: A Guide*. Dordrecht: Kluwer Academic Publishers.

Straižys, V. (1992). *Multicolor Stellar Photometry*. Tucson, AZ: Pachart Publishing House.

Struve, O., Lynds, B. & Pillans, H. (1959). *Elementary Astronomy*. New York: Oxford University Press.

Sumi, T., Udalski, A., Szymański, M. *et al.* (2004). The Optical Gravitational Lensing Experiment: catalogue of stellar proper motions in the OGLE II Galactic bulge fields. *MNRAS*, **348**, 1439–1450.

Taylor, J. H., Manchester, R. N. & Lyne, A. G. (1993). Catalog of 558 pulsars. *ApJS*, **88**, 529–568.

Thompson, G. I., Nandy, K., Jamar, C. *et al.* (1978). *Catalogue of Stellar Ultraviolet Fluxes*. London: Science Research Council.

Thorsett, S. E. (2001). In *Encyclopedia of Astronomy & Astrophysics*, vol. 3, ed. P. Murdin. Bristol: IoP, pp. 2177–2183.

Trimble, V. (2010). A review of An *Introduction to the Theory of Stellar Structure and Evolution*, 2nd edn. *Observatory*, **130**, 185–186.

Trumpler, R. J. (1930). *Lick Observatory Bulletin*, No. 420.

Urry, C. M. (1988). X-ray timing of active galactic nuclei. *Lecture Notes in Physics*, **307**, 257–274.

van Biesbroeck, G. (1963). Star catalogues and charts. In *Basic Astronomical Data*, ed. K. Aa. Strand. Chicago IL: University of Chicago Press, pp. 471–480.

van Duinen, R. J., Aalders, J. W. G., Wesselius, P. R. *et al.* (1975). The ultraviolet experiment onboard the Astronomical Netherlands Satellite – ANS. *A&A*, **39**, 159–163.

van Leeuwen, F. (2007). *Hipparcos, the New Reduction of the Raw Data*. London: Springer.

Verbiest, J. P. W., Lorimer, D. R. & McLaughlin, M. A. (2010). Lutz – Kelker bias in pulsar parallax measurements. *MNRAS*, **405**, 564–572.

Viala, Y. P. (1986). Chemical equilibrium from diffuse to dense interstellar clouds. I: Galactic molecular clouds. *A&AS*, **64**, 391–437.

Wall, J. V. & Jenkins, C. R. (2003). *Practical Statistics for Astronomers*. Cambridge: Cambridge University Press.

Weaver, T. A., Zimmerman, G. B. & Woosley, S. E. (1978). Pre-supernova evolution of massive stars. *ApJ*, **225**, 1021–1029.

Weisberg, J. M., Cordes, J. M., Kuan, B. *et al.* (2004). Arecibo 430 MHz pulsar polarimetry: Faraday rotation measures and morphological classifications. *ApJS*, **150**, 317–341.

Wesselius, P. R., van Duinen, R. J., Aalders, J. W. G. & Kester, D. (1980). Ultraviolet colours of main-sequence stars. *A&A*, **85**, 221–232.

Wesselius, P. R., van Duinen, R. J., de Jong, A. R. W. *et al.* (1982). ANS ultraviolet photometry, catalogue of point sources. *A&AS*, **49**, 427–474.

Weygand, J. M., Ipavich, F. M., Wurz, P. Paquette, J. A. & Bochsler, P. (1999). *Plasma dynamics and diagnostics in the solar transition region and corona: Determination of the argon isotope ratio of the solar wind using SOHO/CELIAS/MTOF*. Proceedings of 8th SOHO Workshop, ESA SP-446, pp. 701–705.

White, V., ed (2008). *British Astronomical Association, London: BAA. Handbook for 2008.*

Wieser, M. E. & Berglund, (2009). Atomic weights of the elements 2007 (IUPAC technical report). *Pure Appl. Chem.*, **81**, 2131–2156.

Wilkins, G. A. (1989). IAU Style Manual, Comm. 5. In *IAU Transactions*, XXB.

Winget, D. E., Nather, R. E., Clemens, J. C. *et al.* (1994). Whole Earth Telescope observations of the DBV white dwarf GD358. *ApJ*, **430**, 839–849.

Wolfe, A. M., Jorgenson, R. A., Robishaw, T., Heiles, C. & Prochaska, J. A. (2008). An 84 μG magnetic field in a galaxy at redshift $z = 0.692$. *Nature*, **455**, 638–640.

Wollard, E. W. & Clemence, G. M. (1966). *Spherical Astronomy*. New York and London: Academic Press.

Wright, E. L. (2006). A cosmology calculator for the World Wide Web. *PASP*, **118**, 1711–1715.

Yang, Y.-G. (2009). BVR observations and period variation of the neglected contact binary V343 Orionis. *PASP*, **121**, 699–707.

Zacharias, N., Urban, S. E., Zacharius, M. I. *et al.* (2000). The first US Naval Observatory CCD astrograph catalog. *AJ*, **120**, 2131–2147.

Index

Printed in the United States
by Baker & Taylor Publisher Services